The Cambridge Manuals of Science and
Literature

THE NATURAL HISTORY OF CLAY

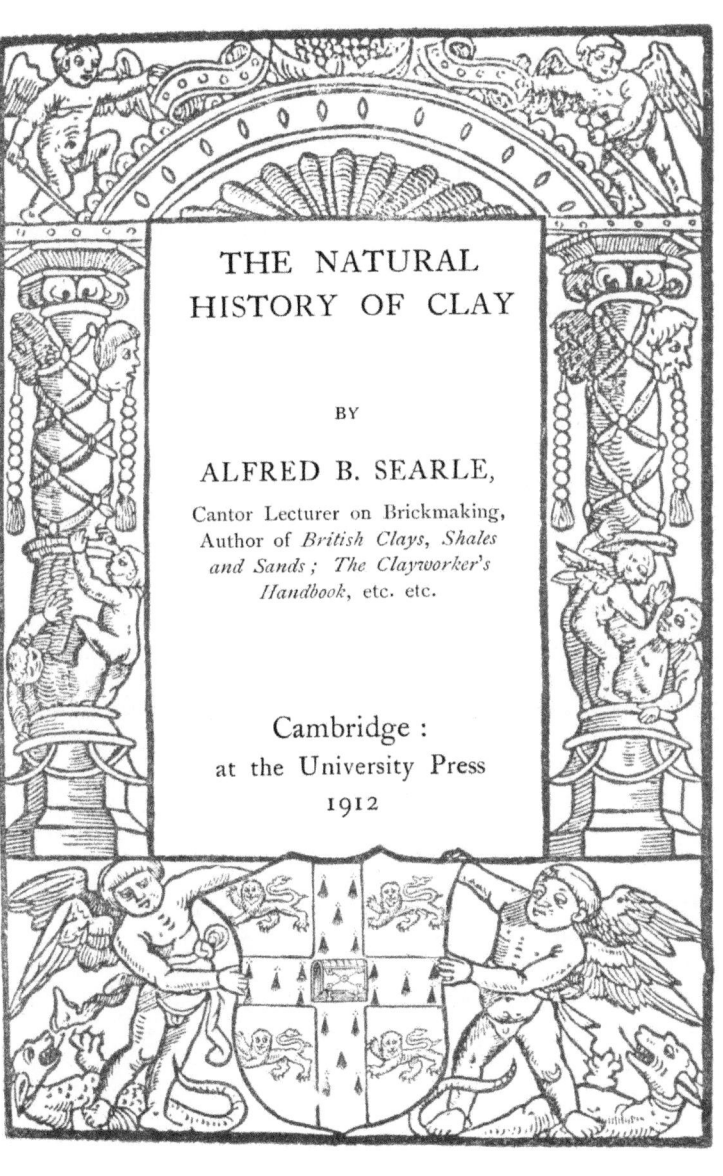

THE NATURAL HISTORY OF CLAY

BY

ALFRED B. SEARLE,

Cantor Lecturer on Brickmaking,
Author of *British Clays, Shales
and Sands; The Clayworker's
Handbook*, etc. etc.

Cambridge :
at the University Press
1912

CAMBRIDGE UNIVERSITY PRESS
Cambridge, New York, Melbourne, Madrid, Cape Town,
Singapore, São Paulo, Delhi, Tokyo, Mexico City

Cambridge University Press
The Edinburgh Building, Cambridge CB2 8RU, UK

Published in the United States of America by Cambridge University Press, New York

www.cambridge.org
Information on this title: www.cambridge.org/9781107698956

First published 1912
First paperback edition 2011

A catalogue record for this publication is available from the British Library

ISBN 978-1-107-69895-6 Paperback

*With the exception of the coat of arms at
the foot, the design on the title page is a
reproduction of one used by the earliest known
Cambridge printer, John Siberch, 1521*

PREFACE

BOTH as raw materials and in the form of pottery, bricks, tiles, terra-cotta and many other articles of use and ornament, clays are amongst the most important rock products. Yet the origin of the substances we know as 'clay,' the processes occurring in its formation and the causes of some of the most important of its characteristics are of such a nature that it is remarkable that its use should have become so extended in the arts and sciences, while we know so little of its properties when in a pure state.

In the following pages an attempt has been made to state in a simple form an outline of our present knowledge of the subject and to indicate the problems which still lie before us.

The experimental solution of these problems is rendered peculiarly difficult by the inertness of the materials at ordinary temperatures and the ease with which the clay molecule appears to break down into its constituent oxides at temperatures approaching red heat or as soon as it begins to react with alkaline or basic materials.

Another serious difficulty is the highly complex nature of that property known as 'plasticity' to which many clays owe their chief value. For many years this has been regarded as an elementary property such as hardness, cohesion or colour, but it is now known to be of so elusive a nature as almost to defy measurement with any degree of accuracy.

The thoroughness with which the methods of physical chemistry have been applied to geological and mineralogical problems during recent years has been of very great assistance to the student of clay problems, as will be seen on studying some of the works mentioned in the short bibliography at the end of the present volume. When the principles of hydrolysis, ionization, mass reaction and reactional velocity have been applied in still further detail to the study of clays, our knowledge of their natural history will increase even more rapidly than it has done during the past few years.

No industry exercises so great a fascination over those engaged in it as do the various branches of clay-working; no other substance offers so many problems of such absorbing interest to the artist, the craftsman, the geologist, the chemist and the general student of nature, whilst the differences in legal opinion as to the nature of clay could themselves occupy a volume far larger than the present one.

A. B. S.

THE WHITE BUILDING,
    SHEFFIELD.
        November 1911.

CONTENTS

CHAP.   PAGE

     Table of clay rocks viii

I   Introduction. The chemical and physical properties
       of clays 1

II   Clay and associated rocks 48

III The origins of clays 70

IV The modes of accumulation of clays 84

V   Some clays of commercial importance . . . 103

VI Clay-substance: theoretical and actual . . . 135

     Bibliography 168

     Index 170

LIST OF ILLUSTRATIONS

FIG.

1 Quartz crystals 9

2 Pyrite 14

3 Marcasite 14

4 Illustrating the structure of a 'clay crumb' . . 24

5 Chart showing rates of drying 27

6 Seger Cones indicating a temperature of 1250° C. . 34

7 Ludwig's Chart 36

8 Coal Measures sequence in North Staffordshire . 55

9 Lias clay being worked for the manufacture of hand-
    made sand-faced roofing tiles 58

10 Oxford clay near Peterborough 60

11 Cliffs of Boulder clay at Filey lying on Calcareous Crag 66

12 China clay pit belonging to the North Cornwall China
    Clay Co. 72

13 Othoclase Felspar 75

14 Illustrating the successive deposition of different strata 90

15 Lacustrine clay at Skipsea 92

16 Clay at Nostel, showing Marine Band . . . 94

17 Kaolinite and Mica 105

18 Mining best Potter's clay in Devonshire . . 111

THE CHIEF CLAY ROCKS (arranged geologically)

Tertiary	Recent (*alluvial clay, silt, brick earths, boulder clay*)
	Pliocene ⎫ Miocene ⎪ (*brick earths, ball clays, coarse pottery clays*) Oligocene ⎬ Eocene ⎭
Secondary	Cretaceous (*cement clays, brick clays*)
	Oolitic (*brick and tile clays*)
	Triassic (*brick, tile and terra-cotta clays*)
Primary	Permian (*brick, tile and flower-pot clays*)
	Carboniferous (*brick clays, fire-clays, ganister*)
	Devonian ⎫ Silurian ⎪ Ordovician ⎬ (*clay schists, slates and clay shales*) Cambrian ⎪ Pre-Cambrian ⎭
	Igneous Rocks occur on several horizons (*china clays and kaolins*)

(In the above Table only the clay-bearing strata are mentioned. The formations named consist chiefly of other rocks in which the clays form strata of variable thickness.)

CHAPTER I

INTRODUCTION. THE CHEMICAL AND PHYSICAL PROPERTIES OF CLAY

THE chief uses of clay have been recognized since the earliest periods of civilization; the ancient Assyrian and Egyptian records contain numerous references to the employment of clay for the manufacture of bricks and for fulling or whitening cloth.

Clays are distributed so widely and in many cases are so readily accessible that their existence and some of their characteristics are known in entirely uncivilized regions. The use of certain white clays as a food, or at any rate as a means of staving off hunger, is common among some tribes of very primitive peoples. The more important uses of clays for building and other purposes are naturally confined to the more civilized nations.

The term *clay* (A.S. *cloeg*; Welsh *clai*; Dutch *kley*) although used in a scientific sense to include a variety of argillaceous earths (Fr. *argile* = clay) used in the manufacture of bricks, tiles, pottery and ceramic products (Gr. *keramos* = potter's earth) generally, is really a word of popular origin and use. Consequently, it is necessary to bear in mind, when considering geological or other problems of a scientific nature, that this term has been incorporated into scientific terminology and that its use in this connection not infrequently leads to confusion. In short, whilst almost every dictionary includes one or more definitions of clay, and most text-books on geology, mineralogy, and allied sciences either attempt a definition or assume the reader's knowledge of one, there is no entirely satisfactory limitation in regard to the substances which may or may not be included under the term.

Clay is a popular term for a variety of substances of very varied origins, of great dissimilarity in their composition and in many of their chemical and physical properties, and differing greatly in almost every conceivable respect. It is commonly supposed that all clays are plastic, but some of the purest china clays are almost devoid of this property and some of the most impure earths used for brick-making possess it in a striking degree. Shales, on the one hand —whilst clearly a variety of clay—are hard and rock-like, requiring to be reduced to powder and very thoroughly mixed with water before they become plastic ; many impure surface deposits, on the other hand, are so highly plastic as to necessitate the addition of other (sandy) materials before they can be used for the manufacture of bricks and tiles.

Attempts have been made to include in the term clay 'all minerals capable of becoming plastic when moistened or mixed with a suitable quantity of water,' but this definition is so wide as to be almost impracticable, and leads to the inclusion of many substances which have no real connection with clays. The limitation of the use of the word 'clay' to the plastic or potentially plastic materials of any single geological epoch is also impracticable, for clays appear to have been deposited in almost every geological period, though there is some difference of opinion as to the time of the formation of certain clays known as *kaolins*.

Clay is not infrequently termed a *mineral*, but this does not apply at all accurately to the many varieties of earths known as 'common clays,' which, together with the 'boulder clays,' contain many minerals and so cannot, as a whole, be included under this term.

Whatever may be the legal significance of the term 'mineral'—which has an important economic bearing on account of minerals being taxed or 'reserved' in some instances where non-minerals (including brick clay) are exempt—there can be no doubt that, scientifically, clay is *not a mineral but a rock*. Whatever mineral (if any) may give the chief characteristic property to the clays as a class must be designated by a special title, for the general term 'clay' will not serve for this purpose. Geologically, the clays are sedimentary rocks, some being unaltered, whilst others—the slates—are notably metamorphosed and can seldom be used for the purposes for which clays are employed.

Most clays may be regarded as a mixture of quartz grains, undecomposed rock débris and various decomposition products of rocks ; if the last-named consists chiefly of certain hydrous alumino-silicates, they may be termed 'clay substance' (see Chapter VI). The imperfections of this statement as a definition are obvious when it is remembered that it may include a mixture of fine sand and clay containing only 30 per cent. of the latter substance.

It is, at the present time, quite impossible to construct an accurate definition of the term 'clay.' The most satisfactory hitherto published defines 'clay' as 'a solid rock composed mainly of hydro-alumino-silicates or alumino-silicic acids, but often containing large proportions of other materials ; the whole possessing the property of becoming plastic when treated with water, and of hardening to a stone-like mass when heated to redness.'

From what has already been written, it will be understood that there is no such entity as a standard clay, for the varieties are almost endless, and the differences between them are sometimes so slight as to be scarcely distinguishable.

A further consideration of this branch of the subject may, however, conveniently be deferred to a subsequent chapter.

The best-known clays are the surface clays, loams and marls, the shales and other sub-surface clays, and the pottery and china clays. The values of these different materials vary enormously, some being almost worthless whilst others are highly valued.

The *surface clays* are chiefly used for the manufacture of bricks and tiles (though some are quite unsuitable for this purpose) and form the soil employed in agriculture in many districts.

The *sub-surface clays* and *shales* are harder, and usually require mechanical treatment before they can be used for brick and terra-cotta manufacture, or for the production of refractory and sanitary articles.

The *pottery and china clays* are usually more free from accessory constituents, and are regarded as the 'purest' clays on the market, though a considerable amount of latitude must be allowed in interpreting the term 'pure.' China clays are by no means pure in the state in which they occur, and require careful treatment before they can be sold.

Further information with regard to the characteristics of certain clays will be found in Chapter v.

THE CHEMICAL PROPERTIES OF CLAY.

The chief constituents of all clays are alumina and silica, the latter being always in excess of the former. These two oxides are, apparently, combined to form a hydro-alumino-silicate or alumino-silicic acid corresponding to the formula $H_4Al_2Si_2O_9$[1], but many clays contain a much larger proportion of silica than is required to form this compound, and other alumino-silicates also occur in them in varying proportions (see Chapters V and VI).

All clays may, apparently, be regarded as consisting of a mixture of one or more hydrous alumino-silicates with free silica and other non-plastic minerals or rock granules, and their chemical properties are largely dependent on the nature and proportion of these accessory ingredients.

The purest forms of clay (china clays and ball clays) approximate to the formula above-mentioned, but others differ widely from it, as will be seen from the analyses on p. 16. The chemical properties of pure clay are described more fully in Chapter VI.

[1] This formula is commonly written $Al_2O_3 2SiO_2 2H_2O$, but although this is a convenient arrangement, it must not be understood to mean that clays contain water in a state of combination similar to that in such substances as washing soda—$Na_2CO_3 24H_2O$, or zinc sulphate crystals—$ZnSO_4 7H_2O$ (see Chapter VI).

Taking china clay, which has been carefully purified by levigation, as representative of the composition of a 'pure' clay, it will be found that the chief impurities in clays are (a) stones, gravel and sand—removable by washing or sifting; (b) felspar, mica and other silicates and free silica—which cannot be completely removed without affecting the clay and (c) lime, magnesia, iron, potash and soda compounds, together with minute quantities of other oxides, all of which appear to be so closely connected with the clay as to be incapable of removal from it by any mechanical methods of purification.

To give a detailed description of the effect of each of the impurities just referred to would necessitate a much larger volume than the present, but a few brief notes on the more important ones are essential to a further consideration of the natural history of clay.

Stones, gravel and *sand* are most noticeable in the boulder clays, but they occur in clays of most geological ages, though in very varying proportions. Sometimes the stones are so large that they may be readily picked out by hand ; in any case the stones, gravel and most of the sand may be removed by mixing the material with a sufficient quantity of water and passing the 'slip' through a fine sieve, or by allowing it to remain stationary for a few moments and then allowing the supernatant liquid to run off

into a settling tank.   Some clays contain sand grains
which are so fine that they cannot be removed in this
manner and the clay must then be washed out by a
stream of water with a velocity not exceeding 2 ft.
per hour.   Even then, the clay so removed may be
found to contain minute grains of silt, much of which
may be removed by a series of sedimentations for
various periods, though a material perfectly free from
non-plastic granules may be unattainable.

Most of the sand found associated with clays is in
the form of fragments of *quartz* crystals (fig. 1), though
it may be composed of irregular particles of other
minerals or of amorphous silica.

Felspar, mica and other adventitious silicates
occur in many natural clays in so fine a state of
division that their removal would be unremunerative.
In addition to this they act as fluxes when the clays
are heated in kilns, binding the less fusible particles
together and forming a far stronger mass than would
otherwise be produced.   Consequently, they are
valuable constituents in clays used for the manufacture of articles in which strength or imperviousness
is important.   If these minerals are present in the
form of particles which are sufficiently large to be
removed by elutriation in the manner described on
the previous page, the purification of the clay is not
difficult.   Usually, however, the most careful treatment fails to remove all these minerals ; their presence

may then be detected by microscopical examination
and by chemical analysis. For most of the purposes
for which clays are used, small proportions of these

Fig. 1.   Quartz crystals, natural size.   (*From Miers'* Mineralogy *by
permission of Macmillan & Co.*)

silicates are unimportant, but where clays of a highly
refractory nature are required; and for most of the
purposes for which china clays (kaolins) are employed,

they must not be present to the extent of more than 5 per cent., smaller proportions being preferable.

Oxides, sulphides, sulphates and *carbonates* of various metals form the third class of impurities in clays. Of these, the most important are calcium oxide (lime), calcium carbonate (chalk and limestone), calcium sulphate (gypsum and selenite), the corresponding magnesia, magnesium carbonate, and sulphate, the various iron oxides, ferrous carbonate and iron sulphides (pyrite and marcasite) (p. 13).

Potash and soda compounds are commonly present as consituents of the felspar, mica, or other silicates present, and need no further description, though small proportions of *soluble salts*—chiefly sodium, potassium, calcium and magnesium sulphates—occur in most clays and may cause a white scum on bricks and terra-cotta made from them.

Lime and magnesia compounds may occur as silicates (varieties of felspar, mica, etc.), but their most important occurrence is as chalk or limestone. *Chalk* is a constant constituent of malms[1] and of many marls, but the latter may contain limestone particles. *Limestone* occurs in many marls and to a smaller extent in other clays. In the boulder clays it frequently forms a large portion of the stony material. If the grains are very small (as in chalk), the lime compounds

[1] A *malm* is a natural mixture of clay and chalk (p. 68).

act as a flux, reducing the heat-resisting power of the clay and increasing the amount of vitrification; they produce in extreme cases a slag-like mass when the clay is intensely heated. If, on the contrary, the grains are larger (as frequently occurs with limestone), they are converted into lime or magnesia when the clay is 'burned' in a kiln, and the lime, on exposure to weather, absorbs moisture (*i.e. slakes*), swells, and may disintegrate the articles made from the clay. Limestone (except when in a very finely divided state) is almost invariably objectionable in clays, but chalk is frequently a valuable constituent.

Chalk is added to clay in the manufacture of malm-bricks to produce a more pleasing colour than would be obtained from the clay alone, to reduce the shrinkage of the clay to convenient limits and, less frequently, to form a more vitrifiable material. Chalk, on heating, combines with iron oxide and clay, forming a white silicate, so that some clays which would, alone, form a red brick, will, if mixed with chalk, form a white one.

Lime compounds have the serious objection of acting as very rapid and powerful fluxes, so that when clays containing them are heated sufficiently to start partial fusion, a very slight additional rise in temperature may easily reduce the whole to a shapeless, slag-like mass. Magnesia compounds act much more slowly in this respect and so are less harmful.

Gypsum—a calcium sulphate—occurs naturally in many sub-surface clays, often in well-defined crystalline masses. It reduces the heat-resisting power of the clays containing it and may, under some conditions, rise to the surface of the articles made from the clay, in the form of a white efflorescence or scum, such as is seen on some brick walls.

Iron compounds are highly important because they exercise a powerful influence on the colour of the burned clays. The red oxide (ferric oxide) is the most useful form in burned clay, but in the raw material ferrous oxide and ferrous carbonate may also occur, though they are converted into the red oxide on heating. The red iron oxide, which is closely related to 'iron rust,' occurs in so finely divided a state that its particles appear to be almost as small as those of the finest clays. Hence attempts to improve the colour of terra-cotta and bricks by the addition of commercial 'iron oxide' are seldom satisfactory, the finest material obtainable being far coarser than that occurring in clays.

It is a curious fact that red iron oxide does not appear to form any compound with the other constituents of clay under ordinary conditions of firing, and although a 'base' and capable of reducing the heat-resisting power of clays, it does not appear to do so as long as the conditions in the kiln are sufficiently oxidizing. It is this which enables red bricks and

other articles to be obtained with remarkable uniformity of colour combined with great physical
strength.   In a reducing atmosphere, on the contrary,
ferrous oxide readily forms and attacks the clay,
forming a dark grey vitreous mass.   If the iron
particles are separated from each other they will,
on reduction, form small slag-like spots, but if they
are in an extremely fine state of division and well
distributed, the brick or other article will become
slightly glossy and of an uniform black-grey tint.
The famous Staffordshire 'blue' bricks owe their
colour to this characteristic; they are not really
'blue' in colour.   The effect of chalk on the colour of
red-burning clays has already been mentioned.

  Iron pyrite (fig. 2) and *marcasite* (fig. 3)—both
of which are forms of iron sulphide—occur in many
clays, particularly those of the Coal Measures.
Mundic is another form of pyrites which resembles
roots or twigs, but when broken show a brassy
fracture.   When in pieces of observable size the
pyrite may be readily distinguished by its resemblance
to polished brass and the marcasite by its tin-white
metallic lustre and both by their characteristic cubic,
root-like and spherical forms; the latter only show
a brass-like sheen when broken.   Even when only
a small proportion of mundic, pyrite or marcasite is
present, it is highly objectionable for several reasons.
In the first place, half the sulphur present is given off

at a dark red heat and is liable to cause troublesome defects on the goods. Secondly, because the remaining sulphur and iron are not readily oxidized, so that there is a great tendency to form slag-spots of ferrous silicate, owing to the iron attacking the clay at the same moment as it parts with its remaining

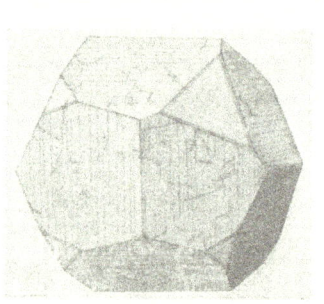

Fig. 2.   Pyrite.   Fig. 3.   Marcasite.
Slightly magnified.

(From Miers' Mineralogy *by permission of Macmillan & Co.)*

sulphur. For this reason, clays containing any iron sulphide seldom burn red, but form products of a buff colour with black spots scattered irregularly over their surface and throughout the mass—an appearance readily observable on most hard-fired firebricks. If chalcopyrite (copper-iron sulphide) is present the spots may be bright green in colour.

Carbon, either free or as hydrocarbons (chiefly vegetable matter) or in other forms, is a constituent of most clays, though seldom reported in analyses. Its presence exercises an important influence in several respects. On heating the clay, with an ample supply of air, the carbonaceous matter may distil off (as shale oil), but more usually it decomposes and burns out leaving pores in the material. If the air-supply is insufficient and the heating is so rapid and intense that vitrification commences before the carbon is all burned away, the pores become filled with the fused ingredients of the clay, air can no longer reach the carbon particles and a black 'core' or heart is produced. Under peculiarly disadvantageous conditions the material may also swell greatly. This is a serious defect in many classes of clay used for brickmaking, and its causes and prevention have been exhaustively studied by Orton and Griffiths (1)[1] but, beyond the brief summary given above, these are beyond the scope of the present work.

Water is an essential constituent of all unburned clays, though the proportion in which it occurs varies within such wide limits that no definite standard can be stated. This water is found in two conditions : (*a*) as moisture or mechanically mixed with the clay particles and (*b*) in a state of chemical combination.

[1] References to original papers, etc. will be found in the appendix.

ANALYSES OF TYPICAL CLAYS

The samples were all dried at 105° C.

Clay...	China Clay	Ball Clay	Fire-clay	Brick Clay	Boulder Clay	Marl
Locality...	Cornwall	Dorset	Yorkshire	Midlands	Lancs.	Suffolk
ULTIMATE ANALYSIS:						
Silica	47·1	49·1	68·9	57·7	63·7	43·7
Alumina	39·1	33·7	19·3	24·3	20·4	15·5
Ferric oxide	·6	1·2	1·0	5·0	3·0	5·2
Titanium oxide	—	·2	1·8	·1	·2	—
Lime	·4	·8	·9	3·7	4·3	16·3
Magnesia	·2	·3	·3	2·5	2·7	2·1
Potash and Soda	·3	2·5	·9	2·8	2·9	·7
Carbon	2·6	4·3	1·8	1·6	·4	1·6
Water	9·3	7·7	4·8	2·0	2·2	2·4
Other Matter	·4	·2	·3	·3	·2	12·5
Total	100·0	100·0	100·0	100·0	100·0	100·0
PROXIMATE ANALYSIS:						
Gravel and Sand	—	8·4	4·6	22·1	23·1	9·2
Silt	—	4·8	9·0	3·1	8·4	16·0
Felspar- and mica-dust	5·2	15·4	10·3	24·3	18·5	8·9
Silica-dust	3·1	4·0	38·0	3·1	12·6	2·0
Free calcium carbonate	—	—	—	2·1	·2	28·4
Free iron oxide and pyrites	·4	·9	·7	4·2	1·6	3·9
'True clay'	91·3	66·5	37·4	41·1	35·6	31·6
Total	100·0	100·0	100·0	100·0	100·0	100·0

For other analyses the books in the Bibliography at the end of the present volume should be consulted, particularly No. 2, *i.e. British Clays, Shales and Sands.*

The amount of mechanically mixed water will naturally vary with the conditions to which the clay has been subjected ; it will be greatest in wet situations and will diminish as the clay is allowed to dry.

The 'combined water,' on the contrary, appears to be a function of the true clay present in the material, and reaches its highest proportions in the china clays and kaolins, which contain approximately 13 per cent. On heating a clay to 105° C. the moisture or mechanically mixed water is evaporated, but the combined water remains unaffected[1] until the temperature is raised to more than 600° C., when it is driven off and the clay is converted into a hard stone-like mass with properties entirely different from those it previously possessed (see Chapter VI).

THE PHYSICAL CHARACTERS OF CLAYS.

The physical characters of clays are of far more interest and importance than their chemical ones, though the two are naturally connected in many ways, and just as the chemical composition of clays is a subject of extreme complexity so is a study of many of their physical properties. Hence only a

[1] Strictly, there is a slight loss at lower temperatures, but it is too small to be important.

few of the more important characteristics can be mentioned here : for further details the reader must consult a larger treatise (2).

Clays are moderately soft, solid bodies, particularly when moistened, and can usually be cut with a knife, though some indurated clays and shales are almost as hard as felspar. Their apparent specific gravity varies greatly, some clays being much more porous than others, but the true specific gravity is usually between 2·5 and 2·65 ; it is similar to that of quartz and slightly lower than that of felspar and mica. Many clays appear to be devoid of structure, but those obtained from a considerable depth below the surface are frequently laminated and have a structure not unlike that of mica. This will be discussed later.

Examined under a microscope, clays are seen to consist of grains of a variety of sizes, the largest of which will usually be found to be composed of adventitious materials such as sand, quartz, felspar, mica, chalk and limestone. The smallest particles —to which clays owe their chief characteristics— are so minute as to make any examination of their shape very difficult, but they are usually composed of minute crystalline plates together with a much larger proportion of apparently amorphous material. The exact nature of both the crystals and the amorphous material is still unknown in spite of many investigations; in the purer clays both forms of substance appear to have the same chemical composition, viz. that of *kaolinite* ($H_4Al_2Si_2O_9$), which the crystalline portion closely resembles.

Clays emit a characteristic yet indefinable odour when moist; the cause of this is very imperfectly understood, though it is not improbably due to decomposing organic matter, as this occurs in most clays.

The colours of freshly-dug clays are extremely varied and range from an almost pure white through all shades of yellow, red and brown to black. The predominating colours are grey or greyish brown and a peculiar yellow characteristic of some surface clays. The natural colour of a clay is no criterion as to its purity, for some of the darkest ball clays produce perfectly white ware on burning, whilst some of the paler clays are useless to the potter on account of the intensity of their colour when they come out of the kiln. The colour of raw clays is largely due to the carbonaceous matter they contain, and as this burns away in the kiln, the final colour of the ware bears no relation whatever to that of the original clay.

The colour of burned ware depends upon the iron compounds in the clay—these producing buff, red, brown or black (usually termed 'blue') articles—on the presence of finely divided calcium carbonate (chalk)

which can destroy the colouring power of iron compounds and produce white ware, and on the treatment
the clay has received in the kiln. A clay which is
white when underfired will usually darken in colour
if heated to vitrification, and one which burns red in
an oxidizing atmosphere may turn blue-grey or black
under reducing conditions. The extent to which the
carbonaceous matter is burned out also determines
the colour of the fired ware.

The presence of adventitious minerals in the clay
may also affect its colour, particularly when fired.

The most obvious feature in a piece of moist clay
is its *plasticity* [1] or ability to alter its shape when
kneaded or put under slight pressure and to retain
its new shape after the pressure has been removed.
It is this property which enables the production of
ornaments, vessels of various kinds, and the many
other articles which are the result of the application
of modelling tools, of moulding or of the action of a
potter's wheel. So long as clay contains a suitable
proportion of moisture it is plastic and may be made
into articles of any desired shape, but if the amount
of moisture in it is reduced or removed completely,
the material is no longer plastic. It may become so,
however, on adding a further suitable quantity of

---

[1] A plastic substance is one with the characteristics of 'a fluid of
so great a viscosity that it does not lose its shape under the influence
of gravitation.'

water and mixing, provided that it has not been excessively heated. If, in the removal of the moisture, the clay has been heated to 600° C. or more, it loses its power of becoming plastic and is converted into a material more closely resembling stone.

The causes of plasticity appear to be somewhat numerous, though there is no generally accepted explanation of this remarkable quality which distinguishes clays from most other substances. It is true that wet sand, soap, wax, lead and some other materials possess a certain amount of plasticity, but not to anything like the same extent as clay.

So far as clays are concerned, their plasticity appears to be connected with the presence of combined water as well as of mechanically mixed water, for if either of these are removed, plasticity—both actual and potential—is destroyed. The part played by water is not, however, completely known, for the many theories which have been advanced only cover some of the conditions and facts.

A number of observers agree that the molecular constitution of clay is peculiar and that it is to this that plasticity is due. Yet the curious fact that the purest clays—the kaolins—are remarkably deficient in plasticity shows that molecular constitution is not, alone, sufficient. Others hold that the remarkably small size of clay particles enables them to pack together more closely than do particles of other materials and to retain around them a film of water which acts partly as a lubricant, facilitating the change of shape of the mass when under pressure, and partly as an adhesive, causing the particles to adhere to each other when the pressure is removed.

Zschokke has laid much emphasis on the importance of molecular attraction between clay and water as a cause of plasticity, and has suggested that the absorption of the water effects a change in the surfaces of the clay particles, giving them a gelatinous nature and enabling them to change their form and yet keep in close contact.

The fact that mica, fluorspar and quartz, when in a sufficiently finely divided state, are also slightly plastic, appears to be opposed to the molecular constitution theory. Smallness of grain undoubtedly has an influence on the plasticity of clay, coarse-grained clays being notably less plastic than others.

Daubrée pointed out that felspar, when ground with water, develops plasticity to a small extent, and Olschewsky carried this observation further and has suggested that clays owe their plasticity to prolonged contact with water during their removal from their place of formation and previous to or during their deposition. A further confirmation of this theory is due to Mellor (3) who showed that on heating china clay with water under very considerable pressure its plasticity was increased and that felspar and some other non-plastic materials developed plasticity under these conditions.

Johnson and Blake (21) supposed that plasticity is due to the clay being composed of extremely minute plates 'bunched together,' a view which was also held by Biedermann and Herzfield, Le Chatelier and others. Olschewsky enlarged this theory by suggesting that the plasticity of certain clays is dependent on the large surface and the interlocking of irregular particles with the plates just mentioned. These theories of interlocking are, however, incomplete, because the tensile strength of clays should accurately represent the plasticity if interlocking were the sole cause. Zschokke has shown that tensile strength is only one factor which must be determined in any attempt to measure plasticity.

E. H. L. Schwarz (35) has suggested that many clays are composed of small globular masses of plates so arranged as to form an open network (fig. 4) which is sufficiently strong not to be destroyed by pressure. In the presence of water and much rubbing the plates are separated and are made to lie flat on each other, thereby giving a plastic and impermeable mass. If this is really the case it would explain the porosity and large surface of some clays and might account for their adsorptive power.

A theory which was first promulgated in 1850 by Way (4), but which has only received detailed attention during the last few years, attributes plasticity to the presence of colloid substances in clay or to the fact that clay particles possess physical characters analogous to those of glue and other colloids. These colloid substances have a submicroscopic or micellian structure; they are web-like, porous and absorb water eagerly. This water may be removed by drying,

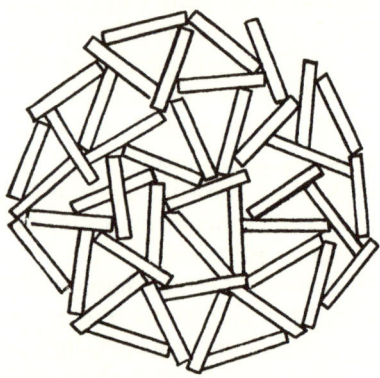

Fig. 4. Illustrating the structure of a ' clay crumb.' (*After Schwarz.*)

only to be re-absorbed on cooling, but if the heating temperature is excessive the structure of the colloids is destroyed. This colloid theory explains many of the facts noted by earlier investigators such as Aron, Bischof, Seger, Olschewsky, etc., but it is not entirely satisfactory, though Rohland (5)—to whom the present prominence of this theory in Europe is largely due—

persistently maintains the contrary. One great objection is the fact that no characteristic *inorganic*
colloid substance has been isolated from pure clay.
It is possible that some of the so-called 'colloidal'
properties of clay may be due to the smallness of its
particles and to their great porosity, as suggested by
Olschewsky.

Despite the present impossibility of producing
a plastic material from artificially prepared colloidal
hydro-alumino-silicates of the same ultimate composition as clay, and the fact that the addition of
colloidal substances does not necessarily increase the
true plasticity of clay, it cannot be denied that the
presence of colloids has an important influence on
it. The addition of starches, glue, gums and similar
substances whilst apparently increasing the plasticity
of clay does not do so in reality. The addition of
1 per cent. of tannin, on the contrary, has been
found by Ries (6) to increase both plasticity and
binding power.

Plasticity appears to be composed of a number of
characteristics so that it is scarcely likely that any
single cause can be assigned to it. On the contrary,
a study of the binding power, tensile strength, ex-
tensibility, adsorption, texture and molecular constitution of clays suggests very strongly that all these
properties are involved in the production of plasticity
and that it is due to the chemical as well as the

physical nature of clay. No clay is entirely colloidal —or it would be elastic and not plastic—but all appear to contain both colloidal and non-colloidal (including plate-like) particles, and it is not improbable that materials in both these states are required, the colloidal matter acting as a cement. Ries (6) has, in fact, pointed out that colloids alone lack cohesiveness and solidity, and a fine mineral aggregate is necessary to change them into a plastic mass resembling clay. The relative proportions of the colloidal material and the sizes of the non-plastic grains will exercise an important influence on all the physical characteristics mentioned above, and therefore on the plasticity.

The manner in which slightly plastic clays become highly plastic in nature is by no means certainly known. It has long been understood that the increase of plasticity is due to changes undergone by the clay during transportation. The most illuminating suggestion is that made by Acheson in 1902, who concluded that it is due to impurities in the water used in transporting the clay or remaining in contact with it during and after its deposition. These impurities may be considered as derived from the washings of forests, and after many experiments with plant extracts Acheson believed the most important substance in this connection to be tannin or gallotannic acid, a dilute solution of which he found

increased the plasticity of china clay by 300 per cent.
From this he further argued that the use of chopped
straw by the Israelites in Egypt in the manufacture
of bricks was unconsciously based on the tannin

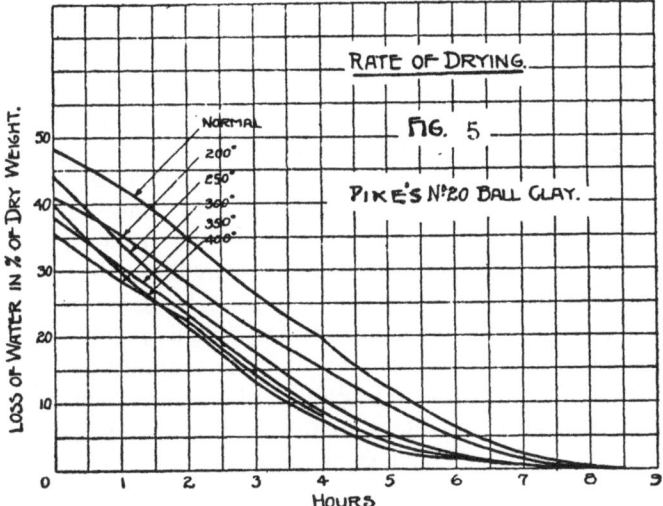

Fig. 5.   Chart showing rates of drying.   (*After Bleininger.*)

content of the straw increasing the plasticity of the
material.

Beadle has stated that 2 per cent. of dissolved
cellulose will increase the plasticity of china clay and
make it equal to that of ordinary clay.

Plasticity is diminished by heating clays, and whilst much of it may be recovered if the temperature has not risen above 400° C. it cannot be completely restored.   Moreover, a clay which has once been heated to a temperature above 100° C. dries in a somewhat different manner to a raw clay.   This is well shown in fig. 5 in which are summarized the results obtained by A. V. Bleininger on a sample of ball clay from Dorset before heating and after portions of it had been heated for 16 hours to 200°, 250°, 300°, 350° and 400° C. respectively.   It is not impossible that if subjected to the influence of water for a sufficiently long time the whole of the plasticity of a heated clay may be restored, providing that the temperature has not been sufficient to cause a destruction of the clay molecule, but as this resumption requires a certain amount of time, Bleininger has proposed to use the reduction in plasticity effected by the heating to enable excessively plastic clays to be worked without the necessity of adding non-plastic material to them.   If any destruction of the clay-molecules has occurred, the plasticity of that portion of the clay can never be restored.

The *binding power* of clays is a characteristic closely connected with plasticity and occasionally confused with it.   All plastic clays have the power of remaining plastic when mixed with materials such as sand, brick-dust ('grog') and other materials which are quite devoid of plasticity. The extent to which a clay can thus bind other materials together into a plastic mass depends, apparently, on the plasticity of the clay itself and on the size and nature of the particles of the added material; the more plastic the clay the larger will be the amount of material it can thus 'bind,' and the finer the latter the more easily will it form a strong material when mixed with a plastic clay.

Rohland (5) has shown that the binding power of clay is not alone due to its cohesion, but that it is closely associated with the colloidal nature of plastic clays: 'fat' clays being those which are highly colloidal, highly plastic and possessing great binding power, whilst 'lean' clays are those deficient in these characteristics. The fact that, as a general rule, the dark coloured clays possess the most binding power, confirms this suggestion, as the dark colour is largely due to organic materials, probably in a colloidal state.

The *shrinkage* which all clays undergo on drying and when heated is another important characteristic. It is due to the fact that as water is removed the solid particles approach closer to each other, the volume of the whole mass being thereby reduced. In a wet piece of clay each particle is surrounded by a film of water, the thickness of which depends on the nature of the clay. As this water evaporates from the surface of the clay its place is taken by water from the interior which rises to the surface by capillary attraction. So long as there is any water between the particles of clay there will be shrinkage when this water is removed, but a stage is eventually reached when the particles of clay are in contact with each other and no more shrinkage can occur. That this cessation of shrinkage may take place before all the water has been removed from the clay is easily understood when it is remembered that whilst the clay particles may be in contact, yet there are still places (pores) where the contact is incomplete, and in these pores water may be retained. The amount of shrinkage clays undergo on drying depends partly on the proportion of water added to them and partly on the sizes of the different particles of clay, sand, etc. present. An average reduction in volume of 12 to 38 per cent. may be regarded as normal, but coarse loams may shrink only 1 per cent. and very finely ground, highly plastic ball clays may shrink as much as 50 per cent., though this is unusual.

As all coagulated colloids, which have absorbed water, shrink on drying, this behaviour of clay appears to confirm the view as to its partially colloidal nature held by some investigators.

When a piece of dry clay is heated sufficiently a further shrinkage (technically known as *kiln shrinkage*) occurs. This begins somewhat below a red heat and increases in rough proportion to the temperature and the duration of the heating. Prolonged heating at a lower temperature will effect the same amount of shrinkage as a short exposure to a higher temperature, but though the greater part of the shrinkage occurs in a comparatively short time, continued heating will be accompanied by a further reduction in volume.

This is due to the fact that clays have no definite melting point, but undergo partial fusion at all temperatures above 950° C. or, in some cases, at even lower ones. As a portion of the material fuses, it fills up the pores in the mass and attacks the unfused material, this process being continued until either the heating is stopped or the whole material is reduced to a viscous slag.

The reduction in the volume of commercial articles made of clay and placed in kilns varies greatly. With bricks, terra-cotta and pottery it must not, usually, exceed 40 per cent. or the warping and cracking which occur will be so great as to make the articles useless. The fineness of the particles exercises an important influence on the kiln shrinkage of a clay, and the latter is frequently reduced in commercial clayworking by adding burned clay ground to a coarse powder to the plastic clay before it is used. Sand is sometimes added for the same purpose, though its more frequent use is to reduce the shrinkage in drying.

Quartz and other forms of free silica expand on heating, so that clays containing them in large quantities shrink very slightly or may even expand.

As clays shrink equally in all directions it is usual to state the contraction in linear instead of volume form. Thus instead of stating that a certain clay when moulded into bricks, dried and burned, shrinks 18 per cent. by volume, it is customary to state that it shrinks ¾ in. per (linear) foot. For many purposes, it is sufficient to regard the linear shrinkage as one-third the volume-shrinkage, but this is not strictly accurate.

The *fusibility* of clays is a characteristic which has been very imperfectly studied. Most clayworkers and investigators employ the term 'fusibility' in a special sense which is apt to be misleading. Owing to the extremely high temperatures to which re-fractory clays can be heated without even losing their shape, it is almost impossible to fuse them completely. In addition to this, clays are not perfectly homogeneous materials and some of their constituents melt at lower temperatures than others. For this reason a clay may show signs of fusion at 1100° C., but it may be heated for some hours at 1800° C. and yet not be completely melted ! Consequently no single 'fusing point' can be stated.

In practice, a suggestion made many years ago by Seger (7) is used ; the clay to be tested is made into a small tetrahedron (fig. 6), heated slowly until it bends over and the point of the test-piece is almost on a level with the base. The temperature at which this occurs is termed the 'fusing point,' though it really only indicates the heat-treatment which is sufficient to soften the material sufficiently to cause it to bend in the manner described. In spite of the apparent crudeness of the test this 'softening point' appears to be fairly constant for most refractory clays.

The bending of a test-piece in this manner is the result of the action of all fluxes[1] in the clay, and as this depends on the size of grain and the duration of the heating above incipient fusion and does not give a direct measure of temperature, nor is the softening effect under one rate of rise in temperature the same as that at another rate. Nevertheless a study of the behaviour of various clays heated simultaneously is valuable and the method forms a convenient means of comparing different materials.

The temperature may be measured by means of a pyrometer, but for the reason just stated it is more convenient and in some respects more accurate to use standard mixtures known as Seger Cones (fig. 6), and to state the softening point in terms of the 'cone' which behaves like the clay being tested. A medium fire-clay will not soften below Seger Cone 26 (1650° C.)

[1] For fluxing materials see p. 8.

and a really good one will have a softening point of cone 34 or 35 (1750° to 1800° C.).

The *refractoriness* of a clay, or its resistance to high temperatures, is an important requirement in bricks required for furnace linings, in crucibles, gas retorts and other articles used in the metallurgical and other industries. The term is much abused and

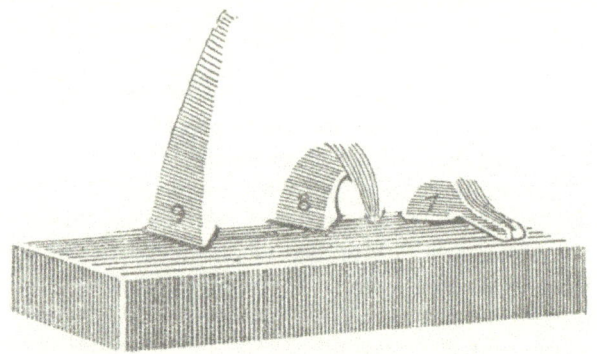

Fig. 6.   Seger Cones indicating a temperature of 1250° C.

is frequently understood to mean resistance to the cutting action of flue gases and flame, the corrosive action of slags, and the strains set up by the repeated changes in temperature. This is unfortunate, for the term refractoriness has a perfectly definite meaning and should be employed exclusively to denote that a given clay is capable of retaining its shape at a given temperature or under given conditions when heated alone and without being subjected to any pressure. In Great Britain there is no officially recognized standard of minimum refractoriness[1], but where one is required the suggested minimum of Seger Cone 26 (1650° C.) made by E. Cramer (8) is usually employed. This is the recognized minimum in Germany for fire-clays, and though objections may be urged against the use of Seger Cones as a standard, equally forcible ones may be brought against making a temperature-scale the basis of measurement. Under present circumstances, however, it is necessary to adopt one or other of these.

Various attempts have been made to ascertain the relationship (if any) between the refractoriness of clays and their chemical composition. If attention is confined strictly to the more refractory clays, some kind of relationship does appear to exist. Thus Richter found that the refractoriness of clay is influenced by certain oxides in the following order : magnesia, lime, ferrous oxide, soda and potash, but this only applies to clays containing less than 3 per cent. of all these oxides. Cramer, in 1895, found that free silica also interfered with the action of these oxides and more recently Ludwig (9) has devised a chart (fig. 7), on the upright sides of which are plotted the equivalents of the lime, magnesia and alkalies,

See *Refractory Clays*, Chapter v.

whilst the silica equivalents are plotted on the horizontal base. In each case the 'molecular formula' of the clay is calculated from its percentage composition, and this 'formula' is reduced so as to have one 'molecule' of alumina, thereby fixing the alumina as a constant and reducing the number of variables to two—the metallic oxides and the silica. Unfortunately

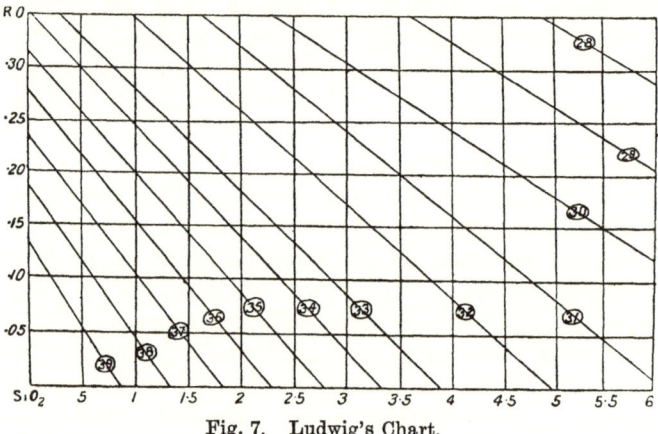

Fig. 7.   Ludwig's Chart.

Ludwig's chart is only applicable to the more refractory clays and cannot be relied upon even for these, though it is extremely useful for comparing clays from identical or similar geological formations.

Attempts to express the refractoriness of clays by means of formulae proving abortive, there only remains the direct test of heating a clay under definite conditions in the manner previously described.

Vitrification is closely connected with the fusibility and refractoriness of clays, and, as a term, indicates the amount of fusion which has occurred under certain conditions of heating. As already mentioned, all clays, on being subjected to a high temperature, undergo partial fusion, the more powerful bases attacking the finest particles of clay and silica, forming molten silicates, and then slowly attacking the more refractory portion; this slow fusion and solution continues until the whole of the material is melted. If the heating is stopped before the fusion has begun, the clay will be porous and comparatively soft, but as more and more material fuses, the mass (on cooling) becomes harder and less porous, as the fused material occupies the pores and sets to a dense, firm glassy mass. The amount of vitrification, or partial fusion, which occurs is, therefore, of great importance in some industries, as by stopping it at an appropriate stage articles of any desired degree of porosity, translucency or strength may be obtained. Thus for common bricks, only sufficient vitrification is permitted to bind the particles firmly together, but in engineering bricks—where much greater strength is required—the vitrification is more complete. Porcelain and earthenware may be similarly distinguished.

The extent to which a given clay will vitrify depends on the amount of fluxing material (metallic compounds, and oxides other than ferric oxide and alumina) it contains, on the smallness of its particles and on the duration and intensity of the heating. Clays containing alkalies and lime compounds vitrify with great rapidity when once the necessary temperature has been reached, so that unless great care is exercised the action will proceed too far and the goods will be warped and twisted or may even form a rough slag. Refractory clays, on the contrary, vitrify more slowly and at much higher temperatures so that accidental overheatings of them are far less common.

The difference between the temperature at which sintering or vitrification occurs and that at which the clay melts completely—usually termed the ' vitrification range'—varies with the nature of the clay. In some cases the clay melts as soon as vitrification becomes noticeable, in others the vitrification occurs at a dull red heat, but the material does not lose its shape until after a prolonged heating at the highest temperature of a firebrick kiln or testing furnace.

Calcareous clays have the melting and sintering points close together, so that it is almost impossible to produce vitrified and impervious ware from them, as they lose their shape too readily. If, however, the difference between the sintering and fusing

temperatures can be enlarged—that is, if the vitrification range can be extended—more impervious ware
can be made. The easiest means of extending the
vitrification range consists in regulating the proportion of large and small particles. The former
increase and the latter diminish the range.

Basic compounds and fluxes cause a lowering of
the melting-point and a shortening of the vitrification
range.

The *porosity* of raw clay is usually of small
importance, but the porosity of fired clay or ware is
often a serious factor in determining the suitability of
certain articles for their intended purposes. In its
natural state, clay does not readily absorb much
water ; on the contrary it becomes pasty and
impervious unless it is disturbed and its texture
destroyed, when it may be mixed with water to form
a paste or, with more water, a thin 'cream' or 'slurry.'

When heated moderately, clay forms a porous
material and, unless the heating is excessive, it will
absorb about one-eighth of its weight of water.
Further heating at a higher temperature reduces its
porosity—the more easily fused material filling some
of the pores—until a stage is reached when the
material is completely vitrified and is no longer
porous.

Porosity may thus be regarded as the opposite of
vitrification ; porous goods being relatively light and

soft whilst vitrified ones are dense and hard. For some purposes, porosity is an important characteristic : for example, building bricks which are moderately porous are preferable to those which are vitrified. The manufacture of porous blocks for the construction of light, sound-proof partitions, etc. has increased rapidly of late. They are made by adding sawdust or other combustible material to the clay. The added substances burn out on firing the goods in a kiln.

Clays which are porous can be dried more readily and with less risk of cracking than those which are more dense. For this reason, some clayworkers mix non-plastic material such as sand or burned clay with their raw material.

The *impermeability* of plastic clay to water is a characteristic which is important for many purposes.

The *absorptive power* of clays is closely related to their porosity so far as pure water is concerned, but if the water contains certain salts in solution a selective absorption occurs, the bases being retained by the clay in such a manner that they cannot be removed by washing. The selective action is known as *adsorption* and is most noticeable in highly plastic clays. Bourry (10) has shown that the slightly plastic china clays only exercise a small power of adsorbing calcium carbonate from solution, but highly plastic clays may adsorb 20 per cent. of it. The alkaline chlorides and sulphates do not appear to be adsorbed in this manner, but the carbonates are readily removed from solution. All calcium and magnesium compounds appear to be adsorbed, though in variable quantities, the reaction being complicated when several soluble salts are present. Ries (6) has found that gallo-tannic acid is adsorbed readily and increases the plasticity of clay.

Ashley (11) has endeavoured to measure the plasticity of clays by determining their adsorption capacity for various aniline dyes, but his untimely decease prevented the investigation being completed. There is reason to suppose that the relation between adsorption and plasticity is extremely close in many clays and that the former may, to an important extent, be used as a measure of the latter. In some clays, however, this relationship does not exist.

Sand and burned clay only show faint adsorption phenomena; felspar shows them to a slight and almost negligible extent and most of the other non-plastic ingredients of clays are non-adsorptive.

Selective adsorption being an important characteristic of colloidal substances, the possession of this power by plastic clays supports the claim that plasticity is due, at least in part, to the presence of colloids.

The addition of small quantities of a solution of certain substances to a stiff clay paste usually reduces its stiffness, and in some cases turns it into a liquid. The alkalies are particularly powerful in this respect and their action may be strikingly illustrated by mixing a few drops of caustic soda with a stiff clay paste. In a few moments the mixture will be sufficiently liquid to pour readily, but it may be rendered quite stiff again by adding sufficient acid to neutralize the alkali previously used. Weber (12) has utilized this characteristic to great advantage in the production of sanitary ware and crucibles for glass-making by a process of casting which he has patented.

The effect of adding water to a dry clay is curious. At first the particles in contact with the water become sticky and plastic, and if the proportion of water added is suitable and the mixing is sufficiently thorough a plastic mass will be produced, the characteristics of which will depend on the nature of the clay used. This process of mixing clay with a limited amount of water is known as 'tempering.' The proportion of water required to make a paste of suitable consistency for modelling appears to be constant for each clay. If, however, a larger proportion of water is added the particles of clay will be separated so widely from each other that they lose their cohesion, and instead of a plastic mass, the material will form a liquid of cream-like consistency. If a piece of stiff clay paste is suspended in a large volume of water without stirring, disintegration will still occur (though a much longer time will be required) and the clay will be deposited as a sediment at the bottom of the vessel. The leaner the clay or the larger the proportion of non-plastic material it contains, the more rapidly will this disintegration take place. A highly plastic clay will become almost impervious and will retain its shape indefinitely.

If a mixture of clay and water in the form of a cream or slurry be allowed to rest, the larger and less plastic particles will settle, but many of the particles of true clay will remain suspended for several hours and some of them for several days. Some particles of clay are so small that it is doubtful if they would ever settle completely unless some coagulant were added, and as they readily pass through all ordinary filtering media it is extremely difficult to collect them in a pure state. These turbid suspensions of clay may be rapidly cleared by the addition of sodium chloride which increases the surface tension of the solution. The fine particles behave in the same way as colloidal substances, *i.e.* as if they possessed an electrostatic charge. Hence the addition of a salt (electrolyte), whose ions annul the opposite charges of the electric double layer assumed by Helmholtz to be present, enables the particles to coagulate in accordance with the ordinary laws of surface tension (14).

Exposure to the action of air and frost has a

marked effect on many clays. When freshly dug
these may be hard and difficult to crush, but after
exposure they break up readily into small fragments.
Clays differ greatly in the extent to which they are
affected by exposure; some are completely disintegrated by standing 48 hours in the open air,
whilst others are scarcely affected by exposure in
bleak places through several years of storm, sunshine
and frost. Usually, however, the effect of a couple
of nights exposure to hard frost will produce a
marked disintegration of the material.

This process of exposure is known as 'weathering'
and its effects are so important that it is employed
whenever possible for clays requiring to be crushed
before use. All clays are rendered more workable
by exposure, but some of them are damaged by the
oxidation of some impurities (*e.g.* pyrites) in them,
though in other clays this very oxidation, if followed
by the leaching action of rain, effects an important
purification of the material.

Weathering appears to have no effect on the
chemical composition of the particles of true clay in
the material, though it may decompose the impurities
present. On the clay itself its action is largely physical
and consists chiefly in separating the particles slightly
from each other, thereby enabling water to penetrate
the material more readily and facilitating the production of a plastic paste. The disintegrating action

of the weather on some 'clays' is so complete that
they require no crushing but can be converted into
a homogeneous paste by simply kneading them with
a suitable proportion of water.

It is possible that on exposure to the heat of the
sun's rays—particularly in tropical climates—some
chemical decomposition of the clay may occur, but
compared with the purely physical action of weathering the amount of such chemical decomposition must
be relatively unimportant in most cases. It may,
however, account for the presence of free silica and
free alumina in some clays.

The action of the weather on rocks, resulting in
the formation of clays, is described in Chapter III.

Heat effects remarkable changes in the physical
character of clays; the most important of these have
already been noted. At a gentle heat, the clay is
dried and retains most of its power of becoming
plastic when moistened; very little, if any, decomposition occurs. At a higher temperature it loses
its 'combined water,' the clay molecule apparently
dissociating, and a hard stony mass—consisting of
particles of free silica and free alumina cemented
together by the more easily fusible impurities
present—is formed. If the heating is continued the
hardness of the material is increased owing to more
molten silicate having been produced from the impurities present, and on cooling, its tensile strength

and resistance to crushing will be found to be enormously greater than those of the original clay. All potential plasticity is destroyed by heating to 700° C. and no method of restoring it has yet been devised. As clays are abundant, this is not a serious disadvantage for the specially desired characteristics of bricks, terra-cotta, pottery and porcelain are all such as to be incompatible with plasticity. The latter is extremely valuable in the shaping of the wares mentioned, but after the manufacture is completed, the destruction of the plasticity is an essential feature of their usefulness.

If the heating is very prolonged or is repeated several times, clays change other of their physical characters and become brittle and liable to crack under sudden changes of temperature. This is partly due to the further fusion (vitrification) which occurs and partly to the formation of crystalline silicates, notably *Sillimanite* (13).

The extent to which clays are ordinarily heated and the conditions under which they are cooled do not usually induce the formation of crystals; the object of the clayworker being to produce a homogeneous mass, the particles of which are securely held together. The result is that burned clay products are usually composed of amorphous particles cemented by a glass-like material formed by the fusion of some of the mineral ingredients of the original substance. The silicates formed are, therefore, in a condition of solid, super-cooled solution in which the tendency to crystallize is restrained by viscosity.

On raising the temperature of firing or on prolonging the heating at the previous maximum temperature the viscosity of the fused portion is diminished and crystallization may then occur. The facility with which crystallization occurs varies greatly with the composition of the fused material, those silicates which are rich in lime and magnesia crystallizing more readily than those containing potash or soda. Vogt has stated that small quantities of alumina promote the formation of a glassy structure, and Morozewicz has shown that a large excess of this substance must be present if crystallization is to occur.

The study of the reactions which occur when clays are heated is, however, extremely complex, not only on account of the variety of substances present, but also on account of the high temperatures at which it is necessary to work, so that for a further consideration of it the reader should consult special treatises on the fusion of silicates. This subject has now become an important branch of physical chemistry.

CHAPTER II

CLAY AND ASSOCIATED ROCKS

CLAY, as already mentioned, is geologically a rock and not a mineral, and belongs to the important group of sedimentary rocks which have been derived from the igneous or primary ones by processes of weathering, suspension in water and subsequent deposition or sedimentation.

Whatever may be the primary origin of clay, its chief occurrence is in geological formations which have undoubtedly been formed by aqueous action. The materials resulting from the exposure of primary rocks to the action of the elements have been carried away by water—often for long distances—and after undergoing various purifications have been deposited where the speed of the water has been sufficiently reduced.

In some cases they have again been transported and re-deposited and not infrequently clay deposits are found which show signs of subsequent immersion at considerable depths and have every appearance of having been subjected to enormous pressures and possibly to high temperatures.

Some clays have only been carried by small streams and for short distances ; these are seldom highly plastic and resemble the lean china clays and kaolins. Others have been carried by rapidly moving rivers and have been discharged into lakes or into the sea ; they have thus undergone a process of gradual purification by elutriation, the sand and other heavier particles being first deposited and the far smaller particles of clay being carried a greater distance towards the centre of the lake or the quieter portions of the ocean. The nature of such deposits will, naturally, differ greatly from each other, the materials at first associated with the clay, or becoming mixed with it at a later stage, exercising an important influence on its texture, composition and properties. If the transporting stream flows through valleys whose sides are formed of limestone, chalk, sandstone or other materials, these will become mixed with the clay, and to so great an extent has the mixing occurred that very few clays occur in a state even approximating to purity. The majority of clays are contaminated with iron oxide, lime compounds and free silica in such a fine state of division that it is impossible to purify them completely without

destroying the nature of the clay. In addition to this
it must be remembered that the land is continually
rising or sinking owing to internal changes in the
interior of the earth, and that these subterranean
changes bring about tilting, folding, overturning and
other secondary changes, which, later, cause a fresh
set of materials to be mixed with the clays. Further
than this, the action of the weather, of rivers and of
the sea never ceases, so that a process of re-mixing
and re-sorting of materials is continuously taking
place, and has been doing so for countless ages. It
is, therefore, a legitimate cause for wonder that such
enormous deposits of clays of so uniform a character
should occur throughout the length and breadth of
Europe, and practically throughout the world. For
although the composition of many of these beds is of
a most highly complex nature, the general properties
such as plasticity, behaviour on heating, etc., remain
remarkably constant over large areas of country, and
the clays of each geological formation are so much
alike in different parts of the world as to be readily
recognized by anyone familiar with the material of
the same formations in this country. Considerable
differences undoubtedly exist, but these are insignificant in comparison with the vastly different circumstances under which the deposits were accumulated.

Leaving the consideration of the modes of formation of the various clay deposits to later chapters

(III and IV), it is convenient here to enumerate some of the chief characteristics of the different clay deposits and their associated rocks. In this connection it is not proposed to enter into minute details, but rather to indicate in broad outline the chief characteristics of the clays from the different deposits. This general view is the more necessary as clay occurs in each main geological division of the sedimentary rocks and in almost every subdivision in various parts of the world.

The **Precambrian, Cambrian, Silurian and Devonian** 'clays' are chiefly in the form of shales or slates, the latter being clays which have undergone a metamorphic change; the latter resulted in the production of a hard and partially crystalline material with but little potential plasticity and therefore of small importance for the ordinary purposes of clay working.

Slates are distinguished from shales by their splitting into thin leaves which are not in the plane of original deposition, but are due to the deposited material being subjected to great lateral pressure. The rearrangement of the particles thus produced has imparted to the material a cleavage quite independent of the original lamination.

The shales in these formations are occasionally soft and friable and are then termed *marls*, but this name is misleading as they contain no appreciable

4—2

proportion of finely divided calcium carbonate as do the true marls[1].

The clays in the **Carboniferous Limestone** are not, as a whole, of much importance, but the occurrence in this formation of pockets of white refractory clays in Staffordshire, North Wales (Mold) and Derbyshire is interesting, especially as these are used for the manufacture of firebricks and furnace linings. These clays are highly silicious and in composition are intermediate between the Yorkshire fireclays and ganister. Their origin is uncertain, but it is generally considered that they have been produced by the action of the weather and streams on the shales and grits of the Coal Measures which formerly occupied the higher ground around them, though Maw (16) states that 'it is scarcely open to question that they are the remnants of the sub-aerial dissolution of the limestone' (see 'Fireclays,' Chapter v).

In the **Upper Carboniferous System** the clays are highly important because of their general refractory nature, though they differ greatly in this respect, some red-burning shales of this formation having no greater power to resist heat than have some of the surface clays.

[1] Readers desiring more detailed information on the occurrence of the clays mentioned in this chapter should refer to the author's *British Clays* (No. 2 in Bibliography).

Those of the Coal Measures are of two main kinds—shales, or laminated rocks which readily split along the planes of deposition, and unstratified underclays. The *shales* usually occur above the seams of coal and are either of lacustrine or marine origin, differences in their fossils and lithological character supporting one origin for some deposits and the other for the remainder. Some of them are fairly uniform in composition, but others vary so greatly in their physical characters, that they are divided by miners into 'binds' or relatively pure shales, 'rock-binds,' or sandy shales, and sandstones. They also vary greatly in thickness in different localities, and whilst they form the main feature in some districts, in others they are replaced by sandstones.

The *underclays* are so called from their usually lying beneath the coal seams. They are not noticeably stratified and vary greatly in character from soft unctuous materials to hard, sandy rocks. In composition they vary enormously, the percentage of silica ranging from 50 per cent., or less, to as high as 97 per cent.

The mode of formation of the underclays is not certainly known. They do not appear to be soils or of terrestrial origin, but according to Arber (24) correspond closely to the black oozes of marine and semi-marine estuarine deposits of tropical swamps, or to the muds surrounding the stumps of trees in the buried forests of our coast-lines. They thus appear to be quite distinct from the shales above them, both in origin and physical characters. The more silicious portions, known as *Ganister*[1], possess comparatively few of the characteristics of clay though used, like all the more refractory clays of the Coal Measures, for all purposes for which fireclay is employed. The term *fireclay* is, in fact, frequently applied to all the refractory deposits in the Coal Measures, without much regard to their composition (see Chapter v).

Valuable Coal Measure clays occur in enormous quantities in Northumberland, Durham, Yorkshire, Nottinghamshire, Derbyshire, Staffordshire, near Stourbridge, in Warwickshire, Shropshire, North and South Wales and South West Scotland. In Ireland, on the contrary, the Coal Measure clays are of little value except in the neighbourhood of Coal Island, co. Tyrone. The position of the 'Sagger Marls' of North Staffordshire (Keele Series and Etruria Marls), relative to the 'Farewell Rock' or Millstone Grit, is shown in fig. 8 in which the horizontal lines represent coal-seams and ironstone veins.

The dissimilarities in the fossils of the Coal Measure clays and shales in the Northern and Southern Hemispheres suggest that there is a

[1] The Dinas rock used in the Vale of Neath (Wales) is an even more silicious material found in the Millstone Grit immediately below the Coal Measures. It is largely employed for firebricks.

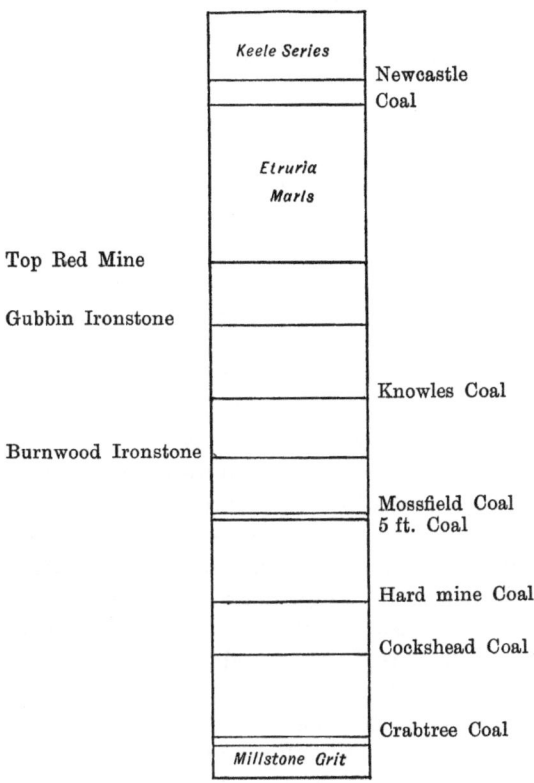

Fig. 8.   Coal Measures sequence in North Staffordshire.

considerable difference in their formation, but the number of clays and shales which have been examined is too small for any accurate conclusion to be drawn.

For many industrial purposes, particularly for the manufacture of refractory goods, the clays and shales of the Carboniferous System are highly important. The less valuable burn to a reddish colour, often spoiled with many grey spots of ferrous silicate derived from the pyrites in the clay, but the purer varieties burn to a delicate primrose or pale buff tint and are amongst the most heat-resisting materials known. The Coal Measure clays of Yorkshire are particularly esteemed for their refractory properties ; for the manufacture of glazed bricks and for blocks for architectural purposes somewhat ambiguously termed 'glazed terra-cotta.'   The inferior qualities are largely used for the manufacture of red engineering bricks, some of them competing successfully with the more widely known ' blue bricks ' of Staffordshire.

The Coal Measure clays of Shropshire are noted for the manufacture of red roofing tiles, especially in the neighbourhood of Broseley.

Agriculturally, the Coal Measure clays are usually poor, but are occasionally of good quality.   The shales produce heavy, cold clays and the yellow subsoil produces soils of a light, hungry character so that the two should, if possible, be mixed together.

Permian clays are of little value except for the manufacture of red building bricks.   The Nottinghamshire Permian clays make excellent roofing tiles, flower pots and red bricks.

Agriculturally, the Permian clays are a free working loam yielding large crops of most of the ordinary farm products.

Triassic clays are of great importance in the Midlands, those upper portions of them known as the Keuper Marls being much used for the manufacture of bricks.

They are specially known amongst clayworkers as the material from which the Midland red bricks of Nottinghamshire and Leicestershire and the Somersetshire tiles are prepared.

Jurassic clays are an important group, of marine origin, occurring in close association with limestone. For this reason they form a valuable source of material for the manufacture of Portland cement, but are of less value to the brick and tile manufacturer.   The Jurassic System contains so large a variety of clays, of such widely different ages and characteristics, that no general description of them can be given in the present volume.

The '*Lias clays*'—the lowest of the Jurassic formation—are chiefly dark, bituminous shales, including the 'alum shales,' and are often seriously contaminated with pyrites and ironstone.   When carefully selected they may be used to advantage in the production of most red articles such as bricks, tiles, chimney pots, etc.   They shrink less in the kiln than do most clays,

Fig. 9.   Lias clay being worked for the manufacture of hand-made sand-faced roofing tiles.   (By courtesy of Messrs Webb Bros. Ltd., Cheltenham.)

and are easily fusible on account of the lime they contain, but on the whole this formation is of great value for the manufacture of the articles just mentioned.

Agriculturally, the Lias clays are laid down for grass, but the lighter soils are useful for arable purposes.

The '*Oolitic clays*,' which are also Jurassic, usually contain limestone in the form of nodules, but are nevertheless important. They form a broad belt above the Lias from Dorset to Yorkshire, and include the blue clays of the Purbeck beds, stiff blue bituminous Kimeridge clays, the irregular, sandy Coral Rag clays, the famous Oxford clay (from which the Peterborough and Fletton bricks are made), the Kellaways blue clay, and the Fuller's Earth deposits.

The '*Kimeridge clays*' are dark, stiff laminated clays, closely resembling gault, and are much used in the West and Midlands for brickmaking. A well-known deposit of this character has long been used at Pickering in Yorkshire, but the most typical deposits are in Huntingdonshire. The Kimeridge clays contain a bituminous shale, or Sapropelic Coal, which evolves a characteristic odour on burning.

Agriculturally, the Kimeridge clays resemble gault and are difficult to work as arable land, though they form first-rate pasturage.

The '*Oxford clays*' are valuable for brickmaking

Fig. 10.   Oxford clay near Peterborough.
(By courtesy of Messrs Ruston, Proctor & Co. Ltd.)

when their use is understood, but to the uninitiated they are very troublesome. Their colour is dark blue or grey and they are usually stiff or somewhat shaly in texture with layers of variable composition. The closely associated Cornbrash (limestone) is a source of trouble unless great care is taken in the selection of the material. 'Oxford clays' are not infrequently traversed by seams of poor coal or by oil-shales.

Agriculturally, Oxford clay is difficult to work and, while much of it is valuable, large portions are poor and cold. When well exposed to frost it is made much lighter, but even then is not very suitable for wheat and autumn sown crops.

The '*Kellaway blue clays*' are often included in the Oxford clays, though they form irregular bands above them and are of fresh-water origin, whilst the Oxford clays are marine deposits. They are chiefly used commercially for domestic firebricks near Oundle and Stamford.

Cretaceous clays occur, as their name implies, in association with chalk. The chief clay in this System is the *gault*, a stiff, black, calcareous clay of marine origin chiefly used for brickmaking. When used alone, gault burns to a reddish colour, due to the iron present, but if, as is more usual, it is mixed with chalk, it burns perfectly white. Some gaults contain sufficient chalk to render the addition of a further quantity unnecessary.

Agriculturally, the Cretaceous clays form good arable soil where they are not too exposed, but they suffer from drought.

The '*Wealden clay*' is a stiff yellowish grey or blue clay extensively used for brickmaking in Kent, Sussex and Surrey. It has been subdivided by geologists into a number of other clays, such as the Wadhurst, Fairlight, etc., but the differences between them lie more in the fossils occurring in them than in the characters of the clays themselves. They are usually contaminated with ironstone, gypsum and some limestone.

Agriculturally, the Wealden clay produces stiff, yellowish soils of a wet and poor character, but sometimes loams of a highly productive nature occur.

The **Tertiary clays** include all those deposited after the Chalk and previous to the close of the Glacial period. They are usually mixed with sand and gravel, and though the deposits are often thin and irregular they are the most generally important of all clays. They vary greatly in character; some, like the London clay, being almost useless unless mixed with other materials, whilst others like the ball clays of Devonshire and Dorset are amongst the purest and most valuable of the plastic clays. The Tertiary clays are divided by geologists into Pliocene, Miocene and Eocene formations; of these the first are commercially unimportant and the second do not exist in Great Britain. At one time the Bovey Tracey clays were considered to be Miocene, but they have recently been classed as Oligocene by Clement Reid.

Agriculturally, the most important of the Tertiary formations is the Eocene, particularly near London, though it is much covered by sand or gravel. The *London clay*, which produces a heavy brown soil, is of slight value, though when properly drained it produces good crops of wheat, beans, and cabbages and other market-garden produce. For this purpose it is greatly improved by the addition of lime and of town manure. The South Hampshire Eocene beds of clay are cold, wet and of small agricultural value.

The Eocene clays are composed of a variety of clays, many of which are only distinguishable by the different fossils they contain. The most important are the Reading clays, the London clay and the Bagshot clays.

The *Reading clays* extend over a considerable area in the South of England and are most valuable near the town from which they derive their name. The best qualities are mottled in a characteristic manner and are particularly suitable for the manufacture of roofing tiles and small terra-cotta—an industry for which Reading is famous.

The *London clay* is always a treacherous material and is best avoided in the manufacture of bricks and other articles except under highly skilled technical advice.

The *Bagshot clays* in Dorsetshire are famous for the ball and pipe clays shipped from Poole, whilst at Bovey Tracey and in several parts of Devonshire equally valuable ball clays are found and are shipped from Teignmouth.

These *ball clays* are of variable composition and colour and require careful selection and testing. They are closely associated with sands, but the lower beds of clay are remarkably stiff, plastic and white-burning. The colour of the raw clay varies from a pale yellow to a dark brown or even to black, but this is little or no criterion of the colour of goods made therefrom, as the colour is due to carbonaceous matters, 4 per cent. or more carbon being usually present.

The 'blue' and 'black' ball clays are the most valued by potters, but the quality is usually ascertained by a burning test.

The value of these ball clays both in Devonshire and Dorset is due to their comparative freedom from iron and alkalies and to their remarkable unctuousness and plasticity. They are, therefore, largely used in the manufacture of all kinds of earthenware of which they form the foundation material.

In composition, ball clays appear to consist chiefly of a hydro-alumino-silicate corresponding to the formula $H_4Al_2Si_2O_9$, and in this they very closely resemble the china clays (kaolins). The latter are, however, but slightly plastic whilst the ball clays are amongst the most plastic clays known.   The china clays are also much more refractory than the ball clays owing to the somewhat larger proportion of alkalies in the latter.

Pipe clays are an inferior quality of ball clay; they contain rather more iron and alkalies and considerably more silica. For this reason they can only be used for cheaper wares where colour is of less importance and where their excessive contraction can be neutralized by the addition of other substances such as flint.

The **Boulder clays** occur in a blanket-like covering of Drift which lies over the greater part of Northern and Central England, and over a considerable portion of Scotland and Ireland.   They are a product of the Ice Age and, whilst varying greatly in character, may usually be distinguished by the occurrence in them of rounded stones and gravel, some of the former bearing clear indications of glacial action.   The boulder clays are largely used for the manufacture of building bricks, but the strata in which they occur are so irregular that very careful supervision of the digging is necessary.   In some localities these clays form beds 12 ft. or more in thickness and relatively free from gravel ; in other districts the clay is interspersed with lenticular deposits of gravel or sand (commonly known as 'pockets'), and if these are mixed with the clay considerable difficulty in manufacture may be experienced.   The total thickness of the drift deposits is often very great, as in the cliffs at Filey (fig. 11) which are 200 ft. high.

The boulder clays—considered apart from the stones, gravel and sandy materials occurring with them—are usually red-burning, stiff and very plastic, but the gravel, sand and crushed stones mixed with

Fig. 11.   Cliffs of Boulder clay at Filey lying on Calcareous Crag.

them in the formation of the material usually render them of medium plasticity.   By careful washing, most boulder clays may be purified sufficiently to enable coarse brown pottery to be made from them.   Clean deposits of sufficient size to be worked without any purification are occasionally found.

Usually, however, the boulder clay formation is somewhat treacherous as it is difficult to ascertain its nature ; boreholes are apt to be quite misleading as the formation is so irregular in character.

Agriculturally, drift or boulder clays are poor soils, but by judicious management and careful mixing they may be made more fertile.   Where it contains chalk—as in Norfolk and Suffolk—boulder drift forms an excellent arable soil.

Pleistocene or Recent clays are amongst the most important brickmaking materials in the South of England.   They are of remarkably varied character, having been derived from a number of other formations.   Usually the deposits are somewhat shallow and irregular in form, but beds of considerable thickness occur in some localities.

Agriculturally, they are of considerable importance.

Most of the **brick earths** used in the south-east of England are of Recent formation, those of the Thames Valley being of special importance in this connection, particularly where they are associated with chalk ; thus forming natural *marls* or enabling artificial marls to be produced.

The brick earths—in the sense in which this term is used in the south—comprise three important types of clay : (*a*) *Plastic clays* not particularly differentiated from those already described, (*b*) *Loams*

or sandy clays which are sufficiently plastic for
satisfactory use, have the advantage of shrinking but
slightly in drying, and aie largely used in the manufacture of red facing bricks and as light soils, and
(c) *Marls* or calcareous clays, used for the production
of light coloured or white bricks, the chalk they
contain combining with any iron compounds present
and, at the same time, reducing the contractility of
the clay.   On burning, they form a cement which
binds the particles into a strong mass.   These are
the 'true marls' or 'malms' composed of clay and
chalk and must not be confused with the so-called
marls of Staffordshire and elsewhere which are almost
free from lime compounds.   There is, at present, no
definition of 'marl' which is quite satisfactory; a
maker of London stock bricks understanding by this
term a clay containing at least 10 per cent. of chalk;
a maker of white Suffolk bricks a material containing
at least twice this amount; an agriculturalist any
soil, not obviously sandy, which will make his clay
land less sticky ; and many geologists any friable
argillaceous earths.   A general consensus of opinion
is, however, being gradually reached that the term
'marl' should be limited, as far as possible, to clays
containing calcium carbonate in a finely divided
state.

Alluvial deposits—which are also of Recent
formation, though still of sufficient age for skeletons

of mammoths to be found in them—are of so variable a nature as to render any brief, general description impossible. Many of them are so contaminated with sand and crushed limestone as to be useless for manufacturing purposes and of small value agriculturally, but others are important in both these respects.

Further details of the occurrence of clays in the various formations described will be found in the *Maps and Memoirs of the Geological Survey* and in the author's *British Clays* (2).

CHAPTER III

THE ORIGINS OF CLAYS

THE terms 'primary' and 'residual' are applied to those clays which are found overlying or in close association with the rocks from which they have been derived, and distinguish them from the 'secondary' or 'transported' clays which have been carried some distance away from their place of origin.

Residual clays may be formed by the simple removal of other materials, the clay remaining behind, as in the decomposition of some argillaceous limestones, in which the calcareous matter has been removed by solution whilst the clay is unaffected. Such a clay is not a primary one as it has probably been derived from some distant source and, having been deposited along with the limestone ooze, has formed an intimate mixture from which the limestone has, at a later geological epoch, been removed in the manner indicated. Residual clays are seldom pure, being often rich in iron compounds, though the white clays of Staffordshire and Derbyshire are highly refractory.

It is seldom necessary to distinguish residual clays from other secondary or transported ones (Chapters II and IV).

Primary clays, on the contrary, have been derived from rocks which have undergone chemical decomposition, one of the products being clay. The most important primary clays are the kaolins, which are derived from the decomposition of felspar, but other primary clays derived from other minerals are known, though less frequently mentioned.

The *kaolins* are primary clays[1] formed by the decomposition of felspar and occur in many parts of the world.   In Great Britain the most important are the china clays found in Devon and Cornwall, which occur in association with the granite from which they have been formed.   The kaolins in Germany are, apparently, of similar origin, though some are derived from porphyry and not from granite ; they are the chief material used in the manufacture of Dresden, Meissen, Berlin and other porcelains. The French kaolins from St Yrieux and Limousin are said by Granger (17) to be derived from gneiss amphibole.   The American kaolins have, according to Ries (6), been chiefly formed from the weathering of pegmatite veins, but the origin of some important

[1] Some kaolins in central Europe appear to have been transported and of secondary origin.

deposits in Texas and Indiana has not yet been fully explained.

Fig. 12.   China clay pit belonging to the North Cornwall China
Clay Co.   (*By courtesy of W. H. Patchell Esq.*)

The corresponding material used by the Chinese for the manufacture of porcelain bears a name which is really that of the place from whence it was originally obtained ; the term *Kao-ling* indicates merely a high ridge. According to Richthofen (18) the rock from which Chinese porcelain is made is not a true kaolin, but is allied to the *jades*. The term 'kaolin' is therefore a misnomer when applied to white-burning, primary clays generally, but its use has become so firmly established as to render it permanent.

Kaolins are seldom found in a sufficiently pure state to be used direct, but must be freed from large amounts of undecomposed rock, quartz, mica, etc., by a process of washing and sedimentation. When purified in this manner, the best qualities of china clay yield, on analysis, alumina, silica and water in the proportions indicated by the formula $H_4Al_2Si_2O_9$ together with about 5 per cent. of mica and other impurities. Some high class commercial kaolins contain over 30 per cent. of mica and 10 per cent. of quartz.

The chief constituents of rocks which take part in the production of kaolins appear to be the felspars, but the natural processes by which these felspars are decomposed are by no means perfectly understood. Some kaolins appear to have been formed by weathering and others by subaerial action. Thus

Collins (19) has stated very emphatically that the kaolinization of Cornish felspar has been chiefly effected by fluorine and other substances rising from below and not by carbonic acid and water acting from above. Ries (6) and other American observers are equally convinced that certain kaolins they have examined are the result of 'weathering.' German and French investigators are divided in their opinions, and Fuchs has found that the Passau (Saxony) kaolin is derived from a special mineral, not unlike a soda-lime felspar deficient in silica, to which he has given the name 'porcelain spar.'

The *félspars* form a class of minerals whose chief characteristic is the combination of an alkaline or alkaline-earth base with silica and alumina. Ortho-clase ($K_2O \, Al_2O_3 \, 6SiO_2$)—the chief potassium felspar—is typical of the whole class. When treated with water under suitable conditions, the felspar appears to become hydrolysed and some of the water enters into combination, the potash being removed by solution. Attempts to effect this decomposition artificially have proved abortive though several investigators appear to have effected it to a limited extent by electrolysis or by heating under great pressure (3).

The effect on felspars of waters containing carbon dioxide in solution has been studied by Forschammer, Vogt, and others, and they have concluded that kaolinization may occur with this agent though it does not appear to be the chief cause in the formation of Cornish china clays.

Fig. 13. Othoclase Felspar, natural size. (*From Miers'* Mineralogy *by permission of Macmillan & Co.*)

The probable effect of fluoric vapours has been studied by Collins (19) who confirmed von Buch's observation that fluorides (particularly lepidolite and tourmaline) are constantly associated with china clay; he found by direct experiment that felspar is decomposed by hydrofluoric acid at the ordinary temperature without the other constituents of the granite in which it occurs being affected. This theory is confirmed by the great depths of the kaolin deposits in Cornwall and in Zettlitz (Bohemia) which appear to be too great to render satisfactory any theory of simple weathering though kaolins in other localities, especially in America, appear to be largely the result of weathering. According to Hickling (36) the product of the action of hydrofluoric acid 'has not the remotest resemblance to china clay.'

Kaolin, when carefully freed from its impurities, as far as this is possible, is peculiarly resistant to the action of water. This resistance may be due to its highly complex constitution, as the simpler hydro-alumino-silicates, such as collyrite, show an acid reaction when ground with water. Rohland (5), therefore, suggests that kaolinization is effected by water first hydrolysing the felspar and forming colloidal silica and sodium or potassium hydroxides which are removed whilst the complex alumino-silicate remains in the form of kaolin. Hickling (36), on the contrary, believes that the action of the weather on felspar produces secondary muscovite—a form of mica—and that this is, later, converted into kaolinite or china clay (fig. 17, p. 105).

The various theories which have been propounded may be summarized into three main classes, and whilst it is probable that any one of them, or any one combination, may be true for a particular kaolin, yet the whole process of kaolinization is so complex and the conditions under which it has occurred appear to be so diverse that it is doubtful if any simple theory can be devised which will satisfactorily meet all cases.

(*a*) The decomposition of the granite, and particularly of the felspar within it, may be ascribed to purely chemical reactions in which the chief agents are water and carbon dioxide.

(*b*) Other substances—possibly of an organic nature and derived from the soil—may have played an important part.

(*c*) Wet steam and hot solutions of fluorine, boron or sulphur compounds may have effected the decomposition.

The recent progress made in the application of the laws of physical chemistry to geological problems is continually throwing fresh light on this interesting subject. Thus, studies of the dissociation pressures and transition points between the anhydrous and the hydrous states of various substances and the effect of water as a powerful agent of decomposition (hydrolysis) have shown that hydration is a characteristic result of decompositions occurring in the upper portions of the earth's crust and not in the lower ones, and that it is usually checked, or even reversed, when the substance is under great pressure. At great depths kaolins and other complex hydrous silicates give place to anhydrous ones such as muscovite, andalusite and staurolite. There is, therefore, good reason to believe that the kaolinization of Cornish felspar has occurred at only moderate depths from the surface and that it has been chiefly produced by the action of water containing acid gases in solution. The acid in the water may have been absorbed from the atmosphere, or it may be due to vapours rising from below through the felspathic material.

In Great Britain, china clay occurs in the form of powdery particles apparently amorphous, but containing some crystals, scattered through a mass of harder rock, the whole being known as china clay rock or 'carclazite.' The softer portions of this china clay rock are known as 'growan' and the china clay in it represents only a small proportion of the whole material.

The finer particles of clay and other materials are removed by treatment with water, whereby one-third to one-eighth of the material is separated. This small proportion is then subjected to further washing and sedimentation in order to obtain the china clay in a state of commercial purity. It will thus be understood that the Cornish china clays are not 'deposits' in the usual acceptation of that term, the soft growan from which they are obtained being almost invariably the result of decomposition *in situ* of some species of felspar in disintegrated granite.

The commercial kaolins of France, Germany, America and China very closely resemble the Cornish china clays in composition, but when used in the manufacture of porcelain they create differences in the finished material which are clearly noticeable, though microscopical examination and chemical analysis, at present, fail to distinguish between them in the raw state on account of their great resistance to ordinary chemical and physical forces.

In addition to the breaking up of felspathic rocks with the formation of china clay or kaolin (kaolinization), other decompositions which occur may result in the formation of clays, and an examination of a considerable number of clays by J. M. van Bemmelen (26) has led him to suppose that several different clay-forming forces have been at work in the production of clays. He classifies these under four heads :

(1) *Kaolinization*, or the decomposition of felspathic and similar rocks by the action of telluric water containing active gases in solution.

(2) *Ordinary weathering* in which the action is largely mechanical, but is accompanied by some

hydrolysis owing to the impurities contained in the water which is an essential factor.

(3) *Lateritic action*—or simple decomposition by heat—which occurs chiefly under tropical conditions, but may also occur in temperate climates, and has for its main product a mixture of free silica and alumina, the latter being in the form of (amorphous) 'laterite.' It may not improbably be a result of the decomposition of the clay molecule similar to that which occurs when china clay is heated, as there is no temperature below which it can be said that china clay does not decompose into free silica and alumina (29).

(4) *Secondary reactions* in which the products of one of the reactions previously described may undergo further changes, as the conversion of amorphous clayite into crystalline kaolinite, or amorphous laterite into crystalline hydrargillite.

WEATHERING.

The action of the forces conveniently classed under the term *weathering* are of two main kinds:

(*a*) The *mechanical grinding* of sandstone, quartzite, limestone, and other rocks, causes an addition of adventitious material to clay, the proportions being sometimes so large as to render it necessary to term the material an argillaceous sand, rather than a sandy clay. Some of these grains of mineral matter are so minute and so resistant to the ordinary chemical reagents as to make it extremely difficult to distinguish them from clay.

(*b*) The *chemical decomposition* due to the action of very dilute solutions. By this means simple silicates are decomposed with the formation of colloidal silica which may either remain in solution or may be deposited in a coagulated form. At the same time, some alumino-silicates will be similarly decomposed into colloidal alumino-silicic acids or clays.

The ultimate results of the action of ordinary weathering on silicate rocks are, therefore, sands and clays, the latter being in some ways quite distinct in their origin and physical properties from the china clays. According to J. M. van Bemmelen (26) such clays also contain an alumino-silicate soluble in boiling hydrochloric acid followed by caustic soda, whereas pure china clays are unaffected by this treatment.

The variety of silicates and other minerals which —in a partially decomposed condition—go to form 'clays' is so great that the complete separation of the smallest particles of them from those of the true clay present has never been accomplished and our knowledge of the mineralogical constitution of many of the best known clays is far from complete.

It is highly probable that the action of water does not cease with the formation of clay, but that it slowly effects an increase in the plasticity of the clay. There thus appear to be at least three kinds of primary clay, viz.:

Kaolinic or *china clays* which are chiefly derived from felspar and can be isolated in a relatively pure state. They are highly refractory, but only slightly plastic.

Epigenic or *colloidal clays* derived from kaolinic clays, as a secondary product, or directly from felspar, mica, augite and other alumino-silicates by 'weathering.' They are usually less refractory and much more plastic than the china clays and contain a large percentage of impurities—sometimes in the form of free silica (sand) or of metallic oxides, carbonates, sulphides, sulphates, silicates, or other compounds. Many so-called secondary clays such as pipe clays, ball clays and fireclays may be of this type, though their origin is difficult to trace owing to their subsequent transportation and deposition.

Lateritic or *highly aluminous clays*, of a highly refractory character, but low plasticity. They are usually somewhat rich in iron oxide which materially affects their plasticity. Unlike the china clays, pure lateritic clays are completely decomposed by hydrochloric acid. Bauxite and some of the highly aluminous clays of the Coal Measures appear to be of this type.

Unfortunately these different types of clay are

extremely difficult to distinguish and in many instances they have become mixed with each other and with other materials during the actions to be described in the next chapter, that it is often almost impossible to decide whether the true clay in a given specimen possessed its characteristics *ab initio* or whether it has gained them since the time when it ceased to be a primary clay.

Secondary clays are those which have been produced by the action of the weather and other natural forces on primary clays, the changes effected being of a physical rather than a chemical nature (see Chapter IV).

The essential constituent of secondary clays has not been positively identified. In so far as it has been isolated it differs from the true clay in the primary clays in several important respects, and until its nature has been more fully investigated great caution must be exercised in assigning a definite name to it. For many purposes the term *pelinite* (p. 149) is convenient, being analogous to the corresponding one used for material in china clays (*clayite*, p. 107). These terms are purely provisional and must be discarded when the true mineralogical identities of the substances they represent have been established.

CHAPTER IV

THE MODES OF ACCUMULATION OF CLAYS

FROM whatever sources clays may have been originally derived, the manner in which they have been accumulated in their present positions is a factor of great importance both in regard to their chemical and physical characters and their suitability for various purposes.

As explained in Chapter III, the china clays or kaolins may usually be regarded as primary clays derived from granitic or other felspathic or felsitic rocks by chemical decomposition. Such clays are found near to their place of origin, are usually obtainable in a comparatively pure state and are generally deficient in plasticity. They may occur in beds of small or great depth, but these are not 'accumulations' in the ordinary meaning of that term.

Residual clays (p. 70) also form a distinct class, as unlike the majority of argillaceous materials they are left behind when other substances are removed, usually by some process of solution. In many cases, however, the residual clays are really secondary in character, having been transported from their place of origin, together with limestone or other minerals, the mixture deposited and subjected to pressure and possibly to heat, whereby a rock-like mass is formed. This mass has then been subjected to the solvent action of water containing carbon dioxide or other substances which dissolve out the bulk of the associated minerals and leave the residual clay behind.

The chief agents in the transport and accumulation of clays are the *air*, in the form of wind; *water*, in the form of rain, streams, rivers, floods, lakes and seas, or in the form of ice and snow as in glaciers and avalanches; *earth-movements* such as the changes wrought by volcanoes, earthquakes and the less clearly marked rising and falling of various portions of the earth's crust which result in folded, twisted, sheared, cracked, inclined, laminated and other strata.

These agents have first moved the clay from its original site and have later deposited it with other materials in the form of strata of widely varying area and thickness, some 'clay' beds being several hundred feet in depth and occupying many square miles in area, whilst others are in the form of thin 'veins' only a few inches thick or in 'pockets' of small area and depth. These deposits have in many places been displaced by subsequent earth-movements and have been overlain by other deposits so as to render them quite inaccessible. Others have been covered by deposits several miles thick ; but the greater part of the covering has since been removed by glacial or other forces, so that clays of practically all geological ages may be found within the relatively small area of Great Britain.

THE TRANSPORTATION OF CLAYS.

By the action of wind or rain, or of rain following frost, the finer of particles clay are removed from their primary site and as the rain drops collect into streamlets, these unite to form streams and rivers and the clay with its associated minerals is carried along by the water. As it travels over other rocks or through valleys composed of sandstone, limestone and other materials, some of these substances are dislodged, broken into fragments of various sizes and with the clay are carried still further. In their journey these materials rub against each other and against the banks and bed of the stream, thereby undergoing a prolonged process of grinding whereby the softer rocks are reduced to very fine sand and silt which becomes, in time, very intimately mixed with the clay. If the velocity of the stream were sufficiently great, the mixed materials—derived from as many sources as there are rocks of the districts through which they have passed—would be discharged into a lake or into the sea. Here the velocity of the water would be so greatly reduced that the materials would gradually settle, the largest and heaviest fragments being first deposited and the finer ones at a greater distance.

With most streams and rivers, however, the velocity of the water is very variable, and a certain amount of deposition therefore occurs along the course, the heavier particles only travelling a short distance, whilst the finer ones are readily transported. If the velocity of the stream increases, these finer particles (which include the clay) may become mixed with other particles of various sizes and the materials thus undergo a series of mixings and partial sortings until they are discharged at the river mouth or are left along its sides by a gradual sinking of the water level. The clay will be carried the whole course of the river, unless it is deposited at some place where the velocity of the water is reduced sufficiently to permit it to settle.

If floods arise, the area affected by the water will be increased. The *alluvial clays* have, apparently, been formed by overflowing streams and rivers, the material in suspension in the water being deposited as the rate of flow diminished. Such alluvial deposits contain a variety of minerals—usually in a very finely divided state—clay, limestone-dust or chalk, and sand being those most usually found.

River-deposited clays, i.e., those which have accumulated along the banks, are characterized by their irregularity in thickness, their variable composition and the extent to which various materials are mixed together. This renders them difficult to work and greatly increases the risks of manufacture as the whole character of a fluviatile clay may change completely in the course of a few yards.

According to the districts traversed by the water, the extent to which the materials have been deposited and re-transported and the fresh materials introduced by earth-movements, river-deposited clays may be (*a*) *plastic* and sufficiently pure to be classified as 'clays,' (*b*) *marls* or clays containing limestone-dust or chalk thoroughly mixed with the clay, and (*c*) *loams* or clays containing so much sand that they may be distinguished by the touch from the clays in class (*a*). Intermediate to these well-defined classes there are numerous mixtures bearing compound names such as sandy loams, sandy marls, argillaceous limestone, calcareous sands, and calcareous arenaceous clays, to which no definite characteristics can be assigned.

To some extent a transportation of clays and associated materials occurs in *lakes*, but the chief processes there are of the nature of sedimentation accompanied by some amount of separation. On the shores of lakes, and to a much larger extent on the sea coasts, extensive erosion followed by transportation occurs continuously, enormous quantities of land being annually removed and deposited in some portion of the ocean bed. The erosion of cliffs and the corresponding formation of sand and pebbles are too well known to need further description. It should, however, be noticed that the clay particles, being much finer, are carried so far away from the shore that only pebbles and sand remain to form the beaches, the finer particles forming 'ocean ooze.'

The action of the *sea* in the transport of rock-materials is more intense than that of rivers, the coasts being worn away by repeated blows from the waves and the pebbles and sand grains the latter contain. The ocean currents carry the materials dislodged by the waves and transport them, sometimes to enormous distances, usually allowing a considerable amount of separation to take place during the transit. In this way they act in a similar manner to rivers and streams.

Glaciers may be regarded as rivers of ice which erode their banks and bed in a manner similar to, but more rapidly than, streams of water. Owing to their much greater viscosity, glaciers are able to carry large boulders as well as gravel, sand and clay, so that the materials transported by them are far more complex in composition and size than are those carried by flowing water.

Separation and Sedimentation.

The clay and other particles having been placed in suspension in water by one or more of the natural forces already mentioned, they soon undergo a process of sorting or separation, previous to their deposition. The power of water for carrying matter in suspension depends largely on its velocity, and when this is

Fig. 14.   Illustrating the successive deposition of different strata.

reduced, as when a river discharges into a lake or sea, the larger and denser particles at once commence to settle, the smaller ones remaining longer in suspension, though if the velocity of the water is reduced sufficiently all the particles will be deposited. Hence, the deposits in lakes (*lacustrine*) and at the mouths of rivers (*estuarine*) increase more or less regularly in fineness according to their distance from the point at which the water enters, the gravel and stones being deposited first, then the coarse sand, next the finer sand and finally the silt and clay (fig. 14). If cross-currents are present, the deposits will, naturally, be made more irregular, and in some cases variations in the flow of the transporting water may cause the coarser particles to be carried further than usual so that they may cover some of the finer deposits previously formed; but as the clay and silt particles are so much finer than sand and gravel they usually travel so far before settling that their deposits are very uniform if the area over which they are spread is sufficiently large. Lake-deposited clays are for this reason more uniform than estuarine beds, whilst beds deposited at considerable depths in the sea and at a great distance from land are still more uniform.

A *lacustrine clay* is usually more persistent and uniform than fluviatile beds though sometimes difficult to distinguish from the latter. Some of the most valuable clay deposits are of lacustrine formation; their comparative purity and great uniformity enabling ware of excellent colour and texture to be produced without much difficulty. Thus the Reading mottled clays of the Hampshire basin, on the outskirts of the London basin and in Northern France are well known for the admirable red bricks, tiles and terra-cotta produced from them. Still purer clays deposited at Bovey Heathfield in Devonshire are also of lacustrine origin, though they differ in many respects from the ordinary lake-deposited clays and are of unusual thickness for deposits formed in this manner.

The greater purity of lacustrine clays, as compared

Fig. 15.   Lacustrine clay at Skipsea.   (By courtesy of T. Sheppard Esq.)

with fluviatile ones, is attributed to the much larger
area over which the deposit is spread,—enabling
variations in the deposits to be much less noticeable
than when a smaller area is covered—and to the very
small velocity of the water in lakes, whereby all the
coarser particles are deposited a considerable distance
away from the clays and silts.

Ries (6) has pointed out that many (American)
lake-clays are of glacial origin, having been laid in
basins or hollows along the margin of the ice-sheet
or in valleys which have been dammed by an accumulation of drift across them. Such clay beds are
usually surface deposits of variable thickness and
frequently impure. Like all lacustrine deposits they
show (though in a more marked degree than in the
older and larger lakes) alternate layers of sand and
clay, though the former are usually too thin to be
noticeable except for their action in enabling the
deposited material to be easily split along the lines
of bedding.

Estuarine deposits partake of the nature of both
fluviatile and marine beds, according to their position relative to the river from which they originate.
They are usually uncertain in character and are often
irregular in composition owing to the variations in
the flow of the water. The Estuarine clays of Great
Britain—with the possible exception of the Jurassic
deposits in Northamptonshire and Lincolnshire—are

Fig. 16.   Clay at Nostel, showing Marine Band.   (By courtesy of T. Sheppard Esq.)

of minor importance, but in some countries they form a valuable source of clay.

Marine clays are, as their name implies, those deposited from sea water. They are frequently found at a considerable distance from the shores of the ocean in which they were laid down, and subsequent risings and fallings of the surface of the earth have so altered the areas occupied by sea water, that a large number of marine deposits now form dry land. Though usually of enormous size and of generally persistent character, marine clay deposits vary considerably in the composition of the material at different depths, as well as in different areas. This is only to be expected from the manner of their deposition, from the varied sources of the material and from the numerous river- and ocean-currents taking part in their formation. For this reason it is generally necessary to mix together portions of the deposit drawn from various depths in order to secure a greater uniformity than would be obtained if a larger area were to be worked to a smaller depth.

The Oxford clay which extends from the centre of England to the centre of France is a typical marine clay.

At the bottom of all oceans at the present day is a deposit, of unknown thickness, of red calcareous clay or *ooze* which is steadily increasing in amount and is thereby forming a fresh marine deposit, though at present its inaccessibility deprives it of all economic value.

It is important not to overlook the enormous part played by variations in the level of the land relative to that of the ocean in past ages. For instance, there is abundant evidence to show that practically the whole of Great Britain has been repeatedly submerged to great depths and has been raised to heights far greater than its present average. These oft-repeated risings and settlings have caused great changes in the nature of the deposited materials so that in the Coal Measures, for example, there are deposits of obviously fresh-water origin sandwiched in between others undoubtedly marine. It can readily be understood, as stated by Arber (24), that if, at a given period, the dry land during the formation of the Coal Measures gradually subsided, it would first be covered with clear water, whilst from those portions of the area which occupied the higher ground the rivers and streams continued to pour into their estuary a large amount of fresh-water material. Later, a stage would be reached when mud of marine origin invaded the area and covered the previous deposits. When, after an indefinitely long period, the ground again rose, fresh-water deposits might again form, and this alternation of marine and fluviatile deposits appears to have been repeated with great frequency during the Carboniferous period.

In the Lower Coal Measures of Yorkshire and Lancashire, Stopes and Watson (23) have shown that the shales forming the roof of the Upper Foot Coal were derived from drifted sediments of marine origin.

PRECIPITATED CLAYS.

If the plasticity of some clays is really due to the colloidal nature of their particles, it is obvious that they must have been formed by a process of co-agulation or precipitation at a distance from the site of the minerals from which they have been derived. According to the 'colloid theory,' felspar and other alumino-silicates are decomposed by ' weathering,' the chief effect of which is the formation (by hydrolysis) of a colloidal solution of ' clay.' This apparently clear solution flows along in the form of a small streamlet, joins other streamlets and continues its journey. So long as it is quite neutral or contains free alkali the solution will remain practically clear, but as soon as acids enter the stream, or are formed in it by the decomposition of organic matter, a coagulation of the colloidal matter will commence and the amount of ' clay' thus thrown out of solution will depend on the amount of such free acid.

If the coagulation or precipitation occurs in still water, the 'clay' will be deposited almost immediately, otherwise it will be carried forward until it reaches

S. C.   7

a place where it can be deposited in the manner already described.

Such precipitated clays need not necessarily be pure, as other substances may be present in colloidal form and may be coagulated at the same time as the clay. In addition to these, the admixture of sand and other minerals present in suspension in the solution may become mixed with the particles during coagulation and be deposited with them.

Clays formed in this manner are extremely difficult to identify on account of the highly complex nature of the reactions occurring in their vicinity both during and subsequent to their formation.

RE-DEPOSITED CLAYS.

Although many clays and other materials have been transported and accumulated in the manner described, the majority of those now available have been subjected to repeated transportation and deposition, owing to the frequent and enormous changes in the relative levels of land and water during the various geological epochs. So far as can be ascertained, it is during these changes of position and the recurrent exposure to air and to water containing various substances in solution, together with the almost incessant grinding which took place during the transportation and deposition, that most secondary clays became plastic. If this is the case, it explains the impossibility of increasing the plasticity of clay by artificial means, at any rate on a large scale.

The simplest of the agents of re-deposition are rain-storms and floods which, forming suddenly, may cause the water of a stream or river to flow with unwonted velocity and so carry away previously formed deposits of various kinds. Clays transported in this way are termed by Ries (6) *colluvial* clays, the term 'diluvial' is generally employed in this country. If these are derived from a primary clay which has not travelled far since it left the original granite from which it was formed, they will usually be white-burning and of only slight plasticity, but if the flood affects materials which have already been re-deposited several times, the colluvial clays may be of almost any imaginable composition. Floods of a different character—due to the subsidence of the land so that it is partially covered with lake- or sea-water, which beats on its shores and erodes it in the manner already described—are also important factors in the transportation of clays.

So far as clays are concerned, the action of the sea is both erosive and depository, though the sedimentation in it being that of the pelagic ooze at great depths the clayey material is quite inaccessible. Under certain conditions, however, the sea may erode land in one area and may return the transported material to the land in another area. The diluvial clay-silt known as *warp* in the valley of the Humber is of this character.

Quite apart from the action of water, however, much denudation, transportation and re-deposition of clays and associated materials has been due to the action of ice in the form of glaciers, though these do not appear to have had much effect in increasing the plasticity of the clays concerned.

Glacially deposited clays are characterised by their heterogeneous composition, some of them containing far more sand than true clay, whilst yet retaining a sufficient amount of plasticity to enable them to be used for rendering embankments impervious and for the manufacture of common bricks, and, occasionally, of coarse pottery; others contain so much sand as to be useless for these purposes. Most glacial deposits contain a considerable proportion of stones and gravel which must be removed before the clay can be used.

The large proportion of adventitious matter is due in great part to the much greater erosive force and carrying power of ice as compared with water, resulting in much larger pieces of material being carried, and as the whole of the ice-borne material is deposited almost simultaneously when the glacier melts, only a very small amount of separation of the material into different grades takes place. The comparative freedom from coarse sand of some glacial clays shows that some sorting does occur, but it is very limited in extent as compared with that wrought in materials which have been exclusively transported by water.

For the manufacture of bricks, tiles and coarse pottery in Yorkshire, Lancashire and some of the more northern counties of Great Britain, glacially deposited clays are of great importance in spite of their irregular composition. They are frequently termed 'boulder clays' or 'drift clays' (p. 65), but in using these or any other terms for clays transported by glacial action it is important that they should not be understood to refer to the whole of the deposited matter. Large 'pockets' of coarse sand and gravel frequently occur in deposits of this character and veins of the same materials are by no means uncommon. The custom of some geologists of referring to the *whole* of a glacial deposit as 'boulder clay' has, in a number of cases, led to serious financial loss to clayworkers who have erroneously assumed that, because some 'boulder clays' are used for brick and tile manufacture, all deposits bearing a similar title would be equally suitable. This difficulty would largely be avoided if, as is now increasingly the case, the term 'drift' or 'glacial deposit' were used for the deposits as a whole, the term 'boulder clay' being restricted to the plastic portions and not including pockets of sand, gravel and other non-plastic materials.

Boulder clays—using this term in the limited sense just mentioned—consist of variable quantities of sand and clay, stones and gravel being generally associated with them. The stones may usually be removed by careful picking, and the gravel by means of a 'clay cleaner' which forces the plastic material through apertures too small to allow the gravel to pass. The plastic material so separated is far from being a pure clay and may contain almost half its weight of sand, the greater part of which is readily separated by washing the material.

Boulder clays, when freed from stones and gravel, are sufficiently plastic to meet the needs of most users, without being so highly plastic and contractile as to necessitate admixture with sand or similar material.

Some boulder clays contain limestone in the form of gravel or as a coarse powder produced by the crushing of larger fragments. These are less suitable for manufacturing purposes as the lime produced when the articles are burned in the kilns is liable to swell and to disintegrate them on exposure.

Owing to their origin and the nature of the impurities they contain, boulder clays are never pure and when burned are irregular in colour and somewhat fusible unless subjected to some process of purification.

CHAPTER V

ALTHOUGH clays occur in deposits of almost all geological periods, many of them are of little or no commercial value. This may be due to their situation or to their composition and other characteristics. Thus, a Coal Measure clay is ordinarily quite inaccessible, and to sink a shaft specially to obtain it may be an unprofitable undertaking; if, however, a shaft is sunk for coal the clays in the neighbourhood of the coal seams are rendered accessible and, usually, a certain amount of such clays is brought to the surface in order to remove it out of the way of the coal miners.

Again, a clay deposit may be so far removed from human habitations as to make it practically valueless, but if, for any reason, the population of the district in which the clay is situated grows sufficiently, the clay may become of considerable value. It not infrequently happens, therefore, that the commercial importance of a clay deposit is one which fluctuates considerably, yet, in spite of this fact, there are certain kinds of clay which are nearly always of some commercial value. The most important of these are the kaolins (china clays), the pottery and stoneware clays, the refractory clays (fireclays), the brick and terracotta clays and shales, and the clays used in the manufacture of Portland cement. The origin and manner in which these clays have been accumulated have been described in the previous chapters ; it now remains to indicate their characteristics from the point of view of their commercial value.

Commercial china clays and kaolins in the United Kingdom are not simple natural products but, in the state in which they are sold commercially, have all been subjected to a careful treatment with water, followed by a process of sedimentation whereby the bulk of the impurities have been removed. According to the extent to which this treatment has been carried out, they will contain 10 per cent. or more mica and quartz, with little or no tourmaline, felspar and undecomposed granite. In some parts of Europe and America, kaolins are found in a state of sufficient purity to need no treatment of this kind unless they are to be used for the very highest class of wares.

Mica is usually the chief impurity as its particles are so small and their density resembles that of the purified china clay more closely than do the other minerals. In commerce the term *china clay* is

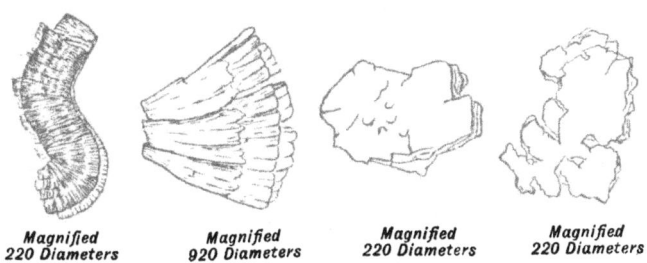

Magnified
220 Diameters
Magnified
920 Diameters
Magnified
220 Diameters
Magnified
220 Diameters

CRYSTALS OF KAOLINITE

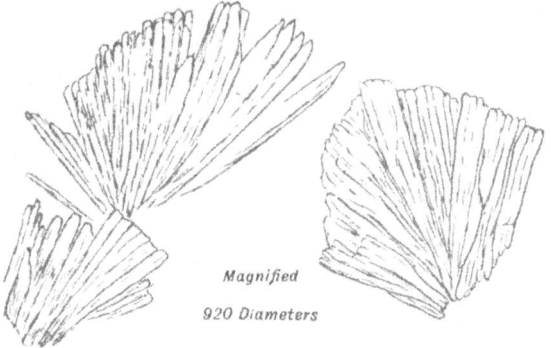

Magnified

920 Diameters

CRYSTALS OF SECONDARY MUSCOVITE.

Fig. 17.   Kaolinite and Mica.   (*After G. Hickling* (36).)

almost invariably used to denote the washed material obtained from the 'china clay rock,' but at the pits the word 'clay' is used indiscriminately for the carclazite (p. 78) and for the material obtained from it. As the term 'kaolin' is used indifferently abroad for the crude 'deposit' and for the purified commercial article, it should be understood that the following information relates solely to the substance as usually sold and not to the crude material.

Commercial china clay or kaolin is a soft white or faintly yellowish substance, easily reduced to an extremely fine powder, which when mixed with twice its weight of water will pass completely through a No. 200 sieve. Its specific gravity is 2·65, but the minuteness and nature of its smallest particles and their character are such that it will remain in suspension in water for several days; it thus appears to possess colloidal properties, at any rate so far as the smaller particles are concerned. It is almost infusible, but shows signs of softening at 1880° C. (Seger Cone 39) or at a somewhat lower temperature, according to the proportion of impurities present. When heated with silica or with various metallic oxides it fuses more readily owing to the formation of silicates.

China clays and kaolins are not appreciably affected by dilute acids, but some specimens are partially decomposed by boiling concentrated hydrochloric acid (26) and all are decomposed by boiling sulphuric acid, the alumina being dissolved and the silica liberated in a form easily soluble in solutions of caustic soda or potash. This has led to the conclusion that some kaolins may have been produced by weathering, as the bulk of true kaolinitic clays (such as Cornish china clay) is not affected by boiling hydrochloric acid (p. 81).

Owing to the exceptional minuteness of its particles, it is extremely difficult to ascertain whether pure china clay or kaolin is crystalline or amorphous. Johnson and Blake (21) found that all the specimens they examined ' consisted largely of hexagonal plates ' and that in most kaolins 'these plates are abundant— evidently constituting the bulk of the substance.' This observation is contrary to the experience of most investigators, the majority of whom have found the bulk of the material to be amorphous and sponge-like, but a small portion of it to consist of hexagonal or rhombic crystals.

Mellor (22) has proposed the name *clayite* for this amorphous material, the crystalline portion being termed *kaolinite* as suggested by Johnson and Blake.

Both kaolinite (crystalline) and clayite (amorphous) yield the same results on analysis and correspond very closely to the formula $H_4Al_2Si_2O_9$ or $Al_2O_3 2SiO_2 2H_2O$, so that it is most probable that they are the same substance in different physical states.

According to Hickling (36) the general impression

that the particles of china clay are amorphous is due to the use of microscopes of insufficient power. With an improved instrument, Hickling claims to have identified the 'amorphous' portion of china clay with crystalline kaolinite, the clay particles (fig. 17) being in the form of irregular, curved, hexagonal prisms or in isolated plates. The former show strong transverse cleavages. The index of refraction and that of double refraction agree with those of Anglesea kaolinite crystals, as does the specific gravity.

In spite of their great purity, commercial china clays and kaolins are almost devoid of plasticity, nor can this property be greatly increased by any artificial treatment. This has led to the conclusion that plasticity is not an essential characteristic of the clayite or kaolinite molecules, but is due to physical causes not shown by any investigation of the chemical composition of the material.

In addition to the specially purified kaolins just described, alkaline kaolins, siliceous kaolins and ferruginous kaolins are obtained from less pure rocks and do not undergo so thorough a treatment with water. Some of these varieties are not improbably derived from transported kaolins, as they occur in Tertiary strata, and so bear some resemblance to the white fireclays on the Carboniferous limestone of Staffordshire, Derbyshire and North Wales, though the latter are far more plastic.

To be of value, a china clay or kaolin must be as white as possible and must be free from more than an insignificant percentage of metallic oxides which will produce a colour when the clay is heated to bright redness. If the material is to be used in the manufacture of paper, paint or ultra-marine, these colour-producing oxides are of less importance providing that the clay is sufficiently white in its commercial state.

The manufacturer of china-ware and porcelain requires china clay or kaolin which, in addition to the foregoing characteristics, shall be highly refractory. It must, therefore, be free from more than about 2 per cent. of lime, magnesia, soda, potash, titanic acid and other fluxes.

It is a mistake to suppose that all white clays of slight plasticity are china clays or kaolins. Some *pipe clays* have these characteristics, but they contain so large a proportion of impurities as to be useless for the purposes for which china clay is employed and are consequently of small value.

Users of china clays and kaolins generally find it necessary to carry out a lengthy series of tests before accepting material from a new source, as such a material may possess characteristics not readily shown by ordinary methods of analysis, but which are sufficiently active to make it useless for certain purposes (see p. 143).

Pottery clays are, as their name implies, those used in the manufacture of pottery, and comprise the china clays already mentioned (p. 104), the ball clays and the less pure clays used in the manufacture of coarse red ware, flower pots, etc.

The *china clays* (p. 104) are not used alone in pottery manufacture as they lack plasticity and cohesion. In the production of china-ware or porcelain they are mixed with a fluxing material such as Cornish stone, pegmatite, or felspar, together with quartz or bone ash. Thus, English china ware is produced from a mixture of approximately equal parts of bone ash, china clay and Cornish stone, whilst felspathic or hard porcelain is made from a mixture of kaolin, felspar and quartz, a little chalk being sometimes added.

The *ball clays* (p. 64) form the basis of most ordinary pottery, though some china clay is usually added in order to produce a whiter ware. Flint is added to reduce the shrinkage—which would otherwise be inconveniently great—and the strength of the finished ware is increased, its texture is rendered closer and its capability of emitting a ringing sound when struck are produced by the inclusion of Cornish stone or felspar in the mixture. Small quantities of cobalt oxide are also added to improve the whiteness in the better classes of ware.

The ball clays are characterised by their remarkably

Fig. 18.   Mining best Potter's clay in Devonshire.   (*Photo by Mr G. Bishop.*)

high plasticity, their fine texture and their freedom from grit. They are by no means so pure as the china clays, and unless carefully selected can only be used for common ware.

The better qualities burn to a vitrified mass of a light brownish tint, but when mixed with the other materials used in earthenware manufacture they should produce a perfectly white ware. The inferior qualities are used for stoneware, drain pipes, etc. It should be noted that the term 'ball clay' is used for clays of widely differing characteristics though all obtained from one geological formation; when ordering it is necessary to state the purpose for which the clay is required or an entirely unsuitable material may be supplied. For the same reason, great care is needed in any endeavour to sell a ball clay from an hitherto unworked deposit.

Coarse pottery clays[1] are usually found near the surface and whilst they may be derived from any geological formation, those most used in England are of Triassic or Permian origin, though some small potteries use material of other periods, including alluvial or surface clays. These clays are closely allied to those used for brickmaking, but are somewhat finer in texture and more plastic. In some cases they are prepared from brick clays by treating the

[1] Coarse pottery has been defined as that made from natural clay without the addition of any material other than sand and water.

latter in a wash-mill, the coarser particles being then
removed, whilst the finer ones, in the state of a slip
or slurry, are run into a settling tank and are there
deposited.

The presence of a considerable proportion of iron
oxide results in the formation of red ware, which is
necessarily of a porous nature, as the fluxes in the
clay are such that they will not permit of its being
heated to complete vitrification without loss of shape.
To render it impervious the ware is covered with
a glaze, usually producing red, brown or black ware
(Rockingham ware).

The *stoneware* or *drain-pipe clays*, are the most
important of the *vitrifiable clays* and owe their value
to the fact that they can be readily used for the
manufacture of impervious ware without the necessity
of employing a glaze. They are, therefore, used in the
manufacture of vessels for holding corrosive liquids
such as acids and other chemicals, for sanitary ap-
pliances, sewerage pipes and in many other instances
where an impervious material is required.

Owing to the lime, magnesia, potash and soda
they contain, the stoneware clays undergo partial
fusion at a much lower temperature than is required by
some of the purer clays. The fused portion fills the
pores or interstices of the material, making—when
cold—a ware of great strength and impermeability.

The chief difficulty experienced in the manufacture

of stoneware is the liability of the articles to twist and warp when heated. For this reason it is necessary to burn them very carefully and to select the clays with circumspection. Some clays are quite unsuitable for this branch of pottery manufacture because of the practical impossibility of producing ware which is correct in shape and is free from warping.

What is required are clays in which the partial fusion will commence at a moderate temperature and will continue until all the pores are filled with the fused material without the remaining ingredients being attacked or corroded sufficiently to cause the ware to lose its shape. As the temperature inside a potter's kiln is continually rising, the great tendency is for the production of fused material to take place at an ever-increasing rate, so that the danger of warping becomes greater as the firing nears completion. Some clays commence to vitrify at a moderate temperature and can be heated through a long range of temperature before an appreciable amount of warping occurs; such clays are said to possess a 'long range of vitrification' (p. 38). In other clays the difference between the temperature at which vitrification commences and that at which loss of shape occurs is only a few degrees; such clays are useless for the manufacture of stoneware, as their vitrification range is too short. It is therefore essential that, for the manufacture of stoneware, a clay should contain a large proportion of refractory material which will form a 'skeleton,' the interstices of which will be filled by the more fusible silicates produced by the firing.

It is generally found that of all the fluxes present in vitrifiable clays, soda and potash compounds—the so-called 'alkalies'—and all lime compounds are the most detrimental, as in association with clay they form a material with a very short range of vitrification. Magnesia, on the contrary, accompanies a long vitrification range.

The clays used in Great Britain for the manufacture of the best stoneware are the Devonshire and Dorset ball clays, the upper portions of these deposits being used for this purpose as they are somewhat less pure than the lower portions used in the manufacture of white ware. For coarser grades of stoneware, clays of other geological formations are employed, especially where the finished ware may be coloured, as the purity of the clay is of less importance. Providing a clay has a sufficiently long vitrification range, a suitable colour when burned, and that it is capable of being readily formed into the desired shapes, its composition and origin are of small importance to the stoneware manufacturer. In actual practice, however, the number of sources of good stoneware clay is distinctly limited, and many manufacturers are thus compelled to add suitable fluxes to refractory clays in order to meet some of their customers' requirements. For this purpose a mixture of fireclay with finely powdered felspar or Cornish stone is used. Chalk—which is a cheaper and more powerful flux—or powdered glass cannot be used as the range of vitrification of the mixture would be too short.

Some manufacturers take the opposite course and add fireclay, flint, or other refractory material to a readily fusible clay. This is satisfactory if the latter clay is relatively low in lime and owes its fusibility to potash, soda or magnesia in the form of mica or felspar. The mica and felspar grains enter so slowly into combination with the clay that a long range of vitrification occurs, whereas with lime, or with some other soda and potash compounds, the combination occurs with great rapidity and the shape of the ware is spoiled.

The **refractory clays** are commonly known as *fireclays* on account of their resistance to heat. The china clays and kaolins are also refractory, but are too expensive and are not sufficiently plastic to be used commercially in the same manner as fireclays, except to a very limited extent, though bricks have been made for many years from the inferior portions of china clay rock at Tregoning Hill in Cornwall.

The geological occurrence of the fireclays of the Coal Measures has already been described on p. 53. In addition, there are the refractory clays occurring in pockets or depressions in the Mountain Limestone of North Wales, Staffordshire, Derbyshire and Ireland, which consist of siliceous clays and sands, the insoluble residue of the local dissolution of the limestone, intermixed with the débris of the overlying Millstone Grit (see p. 54). These clays and sands can be mixed to produce bricks of remarkably low shrinkage, but the pockets are only large enough to enable comparatively small works to be erected and the clays are so irregular both in composition and distribution as to render their use somewhat speculative.

A third type of refractory clay—termed *flint clay* —is used in large quantities in the United States, but is seldom found in Great Britain. When moistened, flint clays do not soften, but remain hard and flint-like with a smooth shell-like fracture. For use they are ground extremely fine, but even then they develop little plasticity. They are considered by Ries (6) to have been formed by solution and re-precipitation of the clay subsequent to its primary formation, in a manner similar to flint. They are somewhat rich in alumina and many contain crystals of pholerite ($Al_2O_3 2SiO_2 3H_2O$).

The Coal Measure fireclays (p. 53)—which are by far the most important—are divided into two sections by the coal seams, those above the coal being shaly and fissile in structure whilst those below (*underclays*) are without any distinct lamination. Both these clays may be equally refractory, but the underclays are those to which the term fireclay is usually applied. The lowest portions are usually more silicious and in some areas are so rich in silica as to be more appropriately termed silica rock or *ganister*. Fireclays may, in fact, be looked upon as a special term for the grey clays of the Coal Measures, interstratified with and generally in close proximity to the seams of coal. They are known locally as *clunches* and *underclays* and were at one time supposed to represent the soil that produced the vegetation from which the coal was formed, but are now considered by many authorities to be of estuarine origin.

It is important to notice that whilst the coals almost invariably occur in association with underclays, some fireclays are found at a considerable distance from coal.

The fireclays of the Coal Measures have a composition varying within comparatively wide limits even in contiguous strata; those chiefly used having an average of 20 to 30 per cent. of alumina and 50 to 70 per cent. of silica. They appear to consist of a mixture of clay and quartz with a small proportion of other minerals, but in some of them a portion of the clay is replaced by halloysite—another hydro-alumino-silicate with the formula

$$H_6Al_2Si_2O_{10} \quad \text{or} \quad Al_2O_3 . 2SiO_2 . 3H_2O.$$

Their grey colour is largely due to vegetable (carbonaceous) matter and to iron compounds. The latter—usually in the form of pyrites—is detrimental to the quality of the goods as it forms a readily fusible slag. Unlike the iron in red-burning clays it can seldom be completely oxidized and so rendered harmless. The fireclays must therefore be carefully selected by the miners.

On the Continent, and to a much smaller extent in Great Britain, refractory articles are made from mixtures of grog or burned fireclay with just sufficient raw clay to form a mass of the required strength. For this purpose a highly plastic, refractory clay is required and the Tertiary ball clays of Devon and Dorset (p. 64) are particularly suitable.

The most important characteristics of a fireclay are that it shall be able to resist any temperature to which it may be exposed and that the articles into which it is made shall not be affected by rapid changes in temperature. Other characteristics of importance in some industries are the resistance to corrosive action of slags and vapours, to cutting and abrasion by dust in flue-gases or by the implements used in cleaning the fires. For those purposes it is necessary that a fireclay should possess high infusibility (p. 32), a low burning shrinkage (p. 29) and a high degree of refractoriness (p. 34), and before it is used these characteristics should be ascertained by means of definite tests, as they cannot be determined by inspection of a sample or from a study of its chemical analysis.

Several grades of fireclay have long been recognized on the Continent and in the United States of America, but the recent Specification of the Institution of Gas Engineers is the only official recognition in Great Britain of definite grades. This specification defines as No. 1 grade a fireclay which shows no signs of fusion when heated to 1670° C. or Cone 30 at the rate of 10° C. per minute, and as No. 2 grade fireclay those which show no signs of fusion when similarly heated to 1580° C. or Cone 26.

It is regarded as a sign of fusion if a test piece with sharp angles loses its angularity after heating to a predetermined temperature (see p. 35).

It is customary to regard as 'fireclay' all clays which, when formed into the shape of a Seger Cone (fig. 6) do not bend on heating slowly until a temperature of 1580° C. (Cone 26) is reached. Any clays comprised within this definition and yet not sufficiently refractory to be of the No. 2 grade just mentioned may be regarded as No. 3 grade fireclays. Many of the last named are well suited for the manufacture of blocks for domestic fireplaces, for glazed bricks and for firebricks not intended to resist furnace temperatures.

To resist sudden changes in temperature the

material must be very porous—the article being
capable of absorbing at least one-sixth of its volume
of water. For this reason it is customary to mix fireclays with a large proportion of non-plastic material
of a somewhat coarse texture, the substance most
generally employed being fireclay which has been
previously burned and then crushed. This material is
known as *grog* or *chamotte* and has the advantage
over other substances of not affecting the composition
of the fireclay to which it is added, whilst greatly
increasing its technical usefulness. The addition of
grog also reduces the shrinkage of the clay during
drying and ensures a sounder article being produced.

The most serious impurities in refractory clays are
lime, magnesia, soda, potash and titanium and their
compounds as they lower the refractoriness of the
material. Iron, in the state of ferric oxide is of less
importance, but pyrites and all ferrous compounds
are particularly objectionable. Pyritic and calcareous
nodules may, to a large extent, be removed by picking,
and by throwing away lumps in which they are seen
to occur. There is, at present, no other means of
removing them.

Fireclays may be ground directly they come from
the mine, but it is usually better to expose them to the
action of the weather as this effects various chemical
and physical changes within the material, which

improves its quality as well as reduces the power required to crush it.

To take full advantage of the refractory qualities of a clay it is necessary to select it with skill, prepare and mould it with care, to burn it slowly and steadily, to finish the heating at a sufficiently high temperature and to cool the ware slowly.

Rapidly heated fireclay is seldom so resistant to heat under commerical conditions as that which has been more steadily fired. Rapid or irregular heating causes an irregular formation and distribution of the fused material during the process of vitrification (p. 37) and so produces goods which are too tender to be durable. It is, therefore, necessary to exercise great care in the firing.

Shales are rocks which have been subjected to considerable pressure subsequent to their deposition and are, consequently, laminated and more readily split in one direction than in others. Some shales are almost entirely composed of silica or calcareous matter, but many others are rich in clay, the term referring to physical structure and not to chemical composition. The clay-shales occur chiefly in the Silurian and Carboniferous formations, the latter being more generally used by clayworkers.

Clay-shales are valued according to (*a*) the proportion of oil which can be distilled from them, those rich in this respect being termed *oil shales ;* (*b*) the colour when burned, as in *brickmaking and terra-cotta shales*; (*c*) the refractoriness, as in *fireclay shales* and (*d*) the facility with which they are decomposed on exposure or on heating and form sulphuric acid as in *alum shales*.

Oil shales contain so much carbonaceous matter that on distillation at a low red heat they yield commercially remunerative quantities of a crude oil termed *shale tar*. In composition they are intermediate between cannel coal and a purely mineral shale. To be of value they should not yield less than 30 gallons of crude oil per ton of shale, with ammonia and illuminating gas as by-products. They are of Silurian, Carboniferous or Oolitic origin, the Kimeridge shale associated with the last-named being very valuable in this respect.

The most important oil shales occur in Scotland.

The *fireclay shales* have already been described on pages 53 and 116.

The *brickmaking shales* are those which are sufficiently rich in clay to form a plastic paste when ground and mixed with water. They can be made into bricks of excellent colour and great strength, but for this purpose require the use of powerful crushing and mixing machinery. They are usually converted into a stiff paste of only moderate plasticity and are then moulded by machinery in specially designed presses, though some firebricks are made from crushed shale mixed into a soft paste with water and afterwards moulded by hand. Some shales, such as the *knotts* at Fletton near Peterborough are not made into a paste, the moist powdered shale being pressed into bricks by very powerful machinery.

Brickmaking shales may be found in any of the older geological formations, though they occur chiefly in the Silurian, Permian, Carboniferous and Jurassic systems. The purer shales of the Coal Measures burn to an agreeable cream or buff colour, the less pure ones and those of the other formations mentioned produce articles of a brick-red or blue-grey colour.

Where the shales are of exceptionally fine grain and their colour when burned is very uniform and of a pleasing tint they are known as *terra-cotta* shales, the red terra-cottas being chiefly made from those occurring in Wales and the buff ones from the lower grade fireclays of the Coal Measures.

Alum shales are characterised by a high proportion of pyrites, which, on roasting, form ferrous sulphate and sulphuric acid. The latter combines with the alumina in the shale and when the roasted ore is extracted with water a solution of iron sulphate and aluminium sulphate is obtained. From this solution (after partial evaporation) alum crystals are obtained by the addition of potassium or ammonium sulphate.

The chief alum shales are those of the Silurian

formation in Scotland and Scandinavia. The Liassic shales of Whitby were at one time an equally important source of alum.

During recent years a large amount of alum has been obtained from other sources or has been made from the lower grade Dorset and Devonshire ball clays by calcining them and then treating them with sulphuric acid. These clays being almost free from iron compounds yield a much purer alum at a lower cost.

Brick clays are those which are not suitable—either from nature or situation—for the manufacture of pottery or porcelain and yet possess sufficient plasticity to enable them to be made into bricks. The term is used somewhat loosely, and geologists not infrequently apply it to clays which are quite unsuitable for brickmaking on account of excessive shrinkage and the absence of any suitable non-plastic medium. Large portions of the 'London clay' are of this nature and can only be regarded as of use to brick- and roofing-tile-manufacturers when the associated Bagshot sands are readily accessible. Similarly, some of the very tough surface clays of the Northern and Midland counties are equally valueless, though designated 'brick clays' in numerous geological and other reports. It is, therefore, necessary to remember that, as ordinarily used, the term 'brick clay' merely indicates a material which appears at first sight to be suitable for brickmaking, but that more detailed investigations are necessary before it can be ascertained whether a material so designated is actually suitable for the purpose.

It is also important to observe that local industrial conditions may be such that a valuable clay may be used for brickmaking because there is a demand for bricks, but not for the other articles for which the clay is equally suitable. For instance, a considerable number of houses in Northumberland and Durham were built of firebricks at a time when it was more profitable to sell these articles for domestic buildings than for furnaces.

In many ways the bricks used for internal structural work form the simplest and most easily manufactured of all articles made from clay. The colour of the finished product is of minor importance and so long as a brick of reasonably accurate shape and of sufficient strength is produced at a cheap rate, little else is expected.

Impurities—unless in excessively large proportions—are of small importance and, indeed, sand may almost be considered an essential constituent of a material to be used for making ordinary bricks. It is, therefore, possible to utilize for this purpose some materials containing so little 'clay' as to make them scarcely fit to be included in this term. So long as the adventitious materials consist chiefly of silica and chalk and the mixture is sufficiently plastic to make strong bricks, it may be used satisfactorily in spite of its low content of clay, but if the so-called 'brick clay' contains limestone, either in large grains or nodules, it will be liable to burst the bricks or to produce unsightly 'blow-holes' on their surfaces. If too much sand or other non-plastic material is present, the resulting bricks will be too weak to be satisfactory.

No brick clay can be regarded as 'safe' if it contains nodules of limestone—unless these can be removed during the preparation of the material— or if the resulting bricks will not show a crushing strength of at least 85 tons per square foot.

The introduction of machinery in place of hand-moulding and of kilns instead of clamps has greatly raised the standard of strength, accuracy in shape and uniformity in colour in many districts, and many builders in the Midlands now expect to sort out from the 'common bricks' purchased, a sufficient number of superior quality to furnish all the 'facing bricks' they require. Apart from this, and in districts where buildings are faced with stone or with bricks of a superior quality, the 'stock' or 'common brick' may be made from almost any clay which will bear drying and heating to redness without shrinking excessively or cracking. A linear shrinkage of 1 in. per foot ($=8\frac{1}{2}$ per cent.) may be regarded as the maximum with most materials used for brickmaking. Clays which shrink more than this must have a suitable quantity of grog, sand, chalk, ashes or other suitable non-plastic material added.

If the clay contains much ferric oxide it will produce red or brown bricks according to the temperature reached in the kiln, but if much chalk is also present (or is added purposely) a combined lime-iron-silicate is produced and the bricks will be white in colour. If only a small percentage of ferric oxide is present a clay will produce buff bricks, which will be spotted with minute black specks or larger masses of a greyish black slag if pyrites is also present or if ferrous silicate has been produced by the reduction of the iron compounds and their subsequent combination with silica.

Further information on brick earths will be found on page 67.

A description of the processes used in the manufacture of bricks being outside the scope of the present work, the reader requiring information on this subject should consult *Modern Brickmaking* (25) or some similar treatise.

Roofing tiles require clays of finer texture than those which may be made into bricks. Stones, if present, must be removed by washing or other treatment, as it is seldom that they can be crushed to a sufficiently fine powder, unless only rough work is required. If sufficiently fine, the clay used for roofing tiles may be precisely the same as that used for bricks and is treated in a similar manner. It must, however, be of such a nature that it will not warp or twist during the burning ; it must, therefore, have a long range of vitrification (p. 38).

Terra-cotta is an Italian term signifying baked earth, but its meaning is now limited to those articles made of clay which are not classed as pottery, such as statues, large vases, pillars, etc., modelled work used in architecture, or for external decoration. Although the distinction cannot be rigidly maintained, articles made of clay may be roughly divided into

(*a*) Pottery (*faience*) and porcelain (glazed),
(*b*) Terra-cotta (unglazed),
(*c*) Bricks and unglazed tiles devoid of decoration.

In this sense, terra-cotta occupies an intermediate position between pottery and bricks, but no satisfactory definition has yet been found for it. Thus, bricks with a modelled or moulded ornament are, strictly, terra-cotta, yet are not so named, and some pottery is unglazed and yet is never classed as terra-cotta, whilst glazed bricks are never regarded as pottery. Again during the past few years, what is termed 'glazed terra-cotta' has been largely used for architectural purposes, yet this is really 'faience.'

Although this overlapping of terms may appear confusing to the reader, it does not cause any

s. c.   9

appreciable amount of inconvenience to the manufacturers or users, as it is not difficult for a practical clay-worker to decide in which of the three classes mentioned a given article should be placed.

Partly on account of the lesser weight, but chiefly in order to reduce the tendency to crack and to facilitate drying and burning, terra-cotta articles are usually made hollow.

It is necessary that clays used in the manufacture of terra-cotta should be of so fine a texture that the finest modelling can be executed. Such clays occur naturally in several geological formations, and some may be prepared from coarser materials by careful washing, whereby the larger grains of sand, stones, etc., are removed. Some shales, when finely ground, make excellent clays for architectural terra-cotta, portions of all the better known fireclay deposits being used for this purpose. It is, however, necessary to use only those shales which are naturally of fine texture, as mechanical grinding cannot effect a sufficient sub-division of the particles of some of the coarser shales.

The finer Triassic 'marls' are also admirable for terra-cotta work, the most famous deposit being the Etruria Marl Series in the Upper Coal Measures near Ruabon.

The most important characteristics required in terra-cotta clays are (*a*) fine texture, or at any rate the ability to yield a fine, dense surface, (b) small shrinkage with little tendency to twist, warp or crack in firing, (c) pleasing and uniform colour when fired, and (d) a sufficient proportion of fluxes to make it resistant to weather without giving a glossy appearance to the finished product.

In large pieces of terra-cotta some irregularity of shape is almost unavoidable, but, if care is taken in the selection and manipulation of the material, this need not be unsightly.

The durability of terra-cotta is largely dependent on the nature of the surface. The most suitable clays, when fired, have a thin 'skin' of vitrified material which is very resistant to climatic influences, and so long as this remains intact the ware will continue in perfect condition. If this 'skin' is removed, rain will penetrate the material and under the influence of frost may cause rapid disintegration.

In the manufacture of very large pieces of terracotta a coarse, porous clay is used for the foundation and interior, and this is covered with the finer clay. By this means a greater resistance to changes in temperature is secured, the drying and the burning of the material in the kiln are facilitated and the risks of damage in manufacture are materially reduced.

Cement clays are those used in the manufacture of Portland cement and of so-called natural cements.

They are largely of an alluvial character and are of two chief classes : (*a*) those which contain chalk or limestone dust and clay in proportions suitable for the manufacture of cement and (*b*) those to which chalk or ground limestone must be added.

They vary in composition from argillaceous limestones containing only a small proportion of clay to almost pure clays.

The manufacture of Portland cement has assumed a great importance and owing to the large amount of investigations made in connection with it, it may be said to represent the chief cement made from argillaceous materials, the others being convenient though crude modifications of it.

The essential constituents are calcium carbonate (introduced in the form of chalk or powdered limestone) and clay, the composition of the naturally occurring materials being modified by the addition of a suitable proportion of one or other of these ingredients. The material is then heated until it undergoes partial fusion and a 'clinker' is formed. This clinker, when ground, forms the cement.

In Kent, the Medway mud is mixed with chalk ; in Sussex, a mixture of gault clay and chalk is employed ; in the Midlands and South Wales, Liassic shales and limestone are used ; in Northumberland a mixture of Kentish chalk and a local clay is preferred, and in Cambridgeshire a special marl lying between

the Chalk and the Greensand is found to be admirable
for the purpose because it contains the ingredients
in almost exactly the required proportions.

For cement manufacture, clays should be as free
as possible from material which, in slip form, will
not pass through a No. 100 sieve, as coarse sand and
other rock débris are practically inert. The proportion
of alumina and iron should be about one-third, but
not more than one-half, that of the silica, and in
countries where the proportion of magnesia in a
cement is limited by standard specifications, it will
be found undesirable to use clays containing more
than 3 per cent. of magnesia and alkalies.

Whilst calcareous clays usually prove the most
convenient in the manufacture of cement, it is by no
means essential to use them, and where a clay almost
free from lime occurs in convenient proximity to
a suitable chalk or limestone deposit an excellent
cement may usually be manufactured.

The 'clays' from which the so-called 'natural' or
'Roman cements' are made by simple calcination and
crushing, usually fuse at a lower temperature than do
the mixtures used for Portland cement, and unless
their composition is accurately adjusted they yield a
product of such variable quality as to be unsuitable
for high class work.

Fuller's earth is a term used to indicate any
earthy material which can be employed for fulling or

degreasing wool and bleaching oil. True fuller's earth is obtained chiefly from the neighbourhood of Reigate, Surrey, Woburn Sands, Bedfordshire and from below the Oolite formation near Bath, but owing to the scarcity of the material and the irregularity of its behaviour, china clay is now largely used for the same purpose. True fuller's earth is much more fusible than the white clays usually substituted for it, and when mixed with water it does not form a plastic paste but falls to powder. As the chief requirement of the fuller is the grease-absorbing power of the material there is no objection to the substitution of other earths of equal efficiency.

Fuller's earth does not appear to be a true clay, though its constitution and mineralogical composition are by no means clearly known. T. J. Porter considers that it is chiefly composed of montmorillonite $(Al_2O_34SiO_2H_2O)$, anauxite, $(2Al_2O_39SiO_26H_2O)$, and chalk with some colloidal silica and a little quartz. It therefore appears to resemble the less pure kaolins, but to contain little or no true clay, though in many respects it behaves in a manner similar to a kaolin of unusually low plasticity.

Other clays of commercial importance, with further details of the ones just mentioned, are described in the author's *British Clays, Shales and Sands* (2).

CHAPTER VI

CLAY SUBSTANCE: THEORETICAL AND ACTUAL

HAVING indicated the origin, modes of accumulation and general characteristics of the numerous materials known as 'clay,' it now remains to ascertain what substance, if any, is contained in all of them and may be regarded as their essential constituent, to which their properties are largely due. Just as the value of an ore is dependent to a very large extent on the proportion of the desired metal which it contains, and just as coal is largely, though not entirely, esteemed in proportion to the percentage of carbon and hydrogen in it, so there may be an essential substance in clays to which they owe the most important of their characteristics.

The proportion of metal in an ore or of hydrocarbon in a coal can be ascertained without serious difficulty by some means of analysis, but with clay the difficulties are so great that, to some extent at least, they must be regarded as being, for the present, insurmountable. This is in no small measure due to the general recognition of all minerals or rocks which become plastic when kneaded with water as 'clays' without much regard being paid to their composition. Consequently materials of the most diverse nature in other respects are termed clays if they are known to become plastic under certain conditions.

There is, in fact, at the present time, no generally accepted definition of clay which distinguishes it from mixtures of clay and sand or other fine mineral particles. The usual geological definitions are so broad as to include many mixtures containing considerably less than half their weight of true clay or they avoid the composition of the material altogether and describe it as a finely divided product of the decomposition of rocks.

Many attempts have been made to avoid this unfortunate position, which is alike unsatisfactory to the geologist, the mineralogist and the chemist as well as to the large number of people engaged in the purchase and use of various clays; and, whilst the end sought has not been reached as completely as is desirable, great progress has been made and much has been accomplished during the last twenty years.

One of the earliest attempts to ascertain whether there is an essential constituent of all clays was made by Seger (7) who used two methods of separating some of the ingredients of natural clays from the remaining constituents. The first of these methods consists in an application of the investigations of

Schulze, Schloesing and Schoene on soils, viz. the removal of the finest particles by elutriation; the second is an extension of the method of Forschammer and Fresenius, viz. the treatment of the material with sulphuric acid.

To the product containing the clay when either of these methods is used Seger gave the name *clay substance*, but the material so separated is by no means pure clay. The term clay substance must, therefore, be confined to the crude product containing the clay together with such other impurities as are in the form of extremely small particles or are soluble in sulphuric acid.

It has not yet been found possible to isolate pure clay from ordinary clays, so that in investigating the nature of what Seger was endeavouring to produce when he obtained the crude clay substance, indirect methods are necessary.

It has long been known that if a sample of ' clay' —using this word in the broadest sense—is rubbed in a considerable quantity of water so as to form a thin slip or slurry, it may readily be divided into a number of fractions each of which will consist of grains of different sizes. This separation may be effected by means of a series of sieves through which the slurry is poured, or the slurry may be caused to flow at a series of different speeds, the material left behind at each rate of speed being kept separate; or, finally, the slurry may be allowed to stand for a few seconds and may then be carefully decanted into another vessel in which it may remain at rest for a somewhat longer period, these times of resting and decantation, if repeated, providing a series of fractions the materials in which are more or less different in their nature.

'Clays' containing a considerable proportion of coarse material are most conveniently separated into a series of fractions by means of sieves, whereby they are divided into (i) stones, (ii) gravel, (iii) coarse sand, (iv) medium sand, (v) fine sand and (vi) a slurry consisting of such small particles that they can no longer be separated by sifting.   If the residues on the sieves are carefully washed free from any adhering fine material and are then dried, they will be found on examination to be quite distinct from anything definable as clay.   They may consist of a considerable variety of minerals or may be almost entirely composed of quartz, but with the possible exception of some shales of great hardness, they are undoubtedly not clay.   This simple process therefore serves to remove a proportion of material which in the case of some 'clays' is very large but in others is insignificant; thus 40 per cent. of sand-like material may be removed from some brick-clays whilst a ball clay used for the manufacture of stoneware or pottery may pass completely through a sieve having 200 meshes per linear inch.

The material which passes through the finest sieve employed will contain all the true clay in the material ; that is to say, the coarser portion will, as already mentioned, be devoid of the ordinary characteristics of clay. At the same time, this very fine material will seldom consist exclusively of clay, but will usually contain a considerable proportion of silt, extremely fine mineral particles and, in the case of calcareous clays, a notable proportion of calcium carbonate in the form of chalk or limestone particles.   Only in the case of the purest clays will the material now under consideration consist entirely of clay, so that it must be again separated into its constituents.   This is best accomplished, as first suggested by Schoene, by exposing the material to the action of a stream of water of definite speed.   H. Seger (7) investigated this method very thoroughly and his recommendations as to the manner in which this separation by elutriation should be carried out remain in use at the present time.   Briefly, all material sufficiently fine to be carried away by a stream of water flowing at the rate of 0·43 in. per minute was found by Seger to include the whole of the clay in the samples he examined, but, as was later pointed out by Bischof, it is not correct to term the whole of this material ' clay substance,' as when examined under the microscope, it contains material which is clearly not clay.

Processes of decantation of the finest material

obtained after elutriation still fail to separate all the non-clay material, and Vogt has found that when the material has been allowed to stand in suspension for nine days some particles of mica are still associated with the clay.

It would thus appear that no process of mechanical separation will serve for a complete purification of a clay; indeed, there are good reasons for supposing that extremely fine particles of quartz and mica render physical characteristics an uncertain means of accurately distinguishing clays from other rock dust.

When chemical methods of investigation are employed the problem is not materially altered, nor is its solution fully attained. It is, of course, obvious that any chemical method should be applied to the product obtained by treating the raw material mechanically as above described, for to do otherwise is to create needless confusion. Yet by far the greater number of published analyses of 'clays' report the ultimate composition of the whole material, no attempt being made to show how much of the various constituents is in the form of sand, stones or other coarse particles of an entirely non-argillaceous character.

If the particles of a 'clay' which are sufficiently small to be carried away by a stream of water with a velocity of only 0·43 in. per minute are analysed, it will be found that their composition will vary according to the origin of the clay and the subsequent treatment to which it has been subjected during its transport and deposition. If the clay is fairly free from calcareous material and is of a white-burning nature it may be found to have a composition like china clays. Red-burning clays, on the contrary, will vary greatly in composition, so that it becomes difficult to find any close analogy between these kinds of clay. This difference is partly due to the extremely fine state of division in which ferric oxide occurs in clays, the particles of this material corresponding in minuteness to those of the purest clays and so being inseparable by any mechanical process.

In 1876 H. Seger (7) published what he termed a method of 'rational analysis,' which consisted in treating the clay with boiling sulphuric acid followed by a treatment with caustic soda. He found that the purer china clays (kaolins) and ball clays were made soluble by this means and that felspar, mica and quartz were to a large extent unaffected. Later investigators have found that this method is only applicable to a limited extent and that its indications are only reliable when applied to the clays just named, but the principle introduced by Seger has proved invaluable in increasing our knowledge of the composition of clays. By means of this so-called rational analysis Seger found that the purer clays yielded results of remarkable similarity and uniformity, the material entering into solution having a composition agreeing very closely with the formula $Al_2O_32SiO_22H_2O$ which is generally recognized as that of the chief constituent or constituents of china clay (kaolin) and the purer ball clays. This crude substance, obtainable from a large number of clays by the treatment just described, was named *clay substance* by Seger, who regarded it as the essential constituent of all clays.

Red-burning clays when similarly treated do not yield so uniform a product, and the ferric oxide entering into solution makes the results very discordant. Moreover, even with the china clays or kaolins a small proportion of alkalies, lime and other oxides enter into solution and a number of minerals analogous to clay, but quite distinct from it, are also decomposed and dissolved. For these reasons the 'rational analysis' has been found insufficient; it is now considered necessary to make an analysis of the portion rendered soluble by treatment with sulphuric acid in order to ascertain what other ingredients it may contain in addition to the true clay present.

As the china clays (kaolins) and ball clays on very careful elutriation all yield a product of the same ultimate composition, viz. 39 per cent. of alumina, 46 per cent. of silica, 13 per cent. of water, and 2 per cent. of other oxides, they are generally regarded as consisting of practically pure clay with a variable

amount of impurities. Many years ago Fresenius
suggested that these non-clayey constituents of clays
should be calculated into the minerals to which they
appeared likely to correspond so as to obtain a result
similar to that obtained by Seger without the dis-
advantages of the treatment with sulphuric acid and
as supplementary to such treatment in the case of
red-burning and some other clays. More recent
investigators have found that if a careful microscopic
examination of the clay is made the results of
estimating the composition from the proportion of the
different minerals recognizable under the microscope
and by calculation from the analysis of the material
agree very closely and are, as Bischof (28) and, more
recently, Mellor have pointed out, more reliable than
the 'rational analysis' in the case of impure clays. If
care is taken to make a microscopical examination
identifying the chief impurities present the calculation
from the analysis may usually be accepted as sufficiently
accurate, but it is very unsatisfactory to assume, as
some chemists do, that the alkalies and lime in the
clay are all in the form of felspar and that the silica
remaining in excess of that required to combine with
the alkalies, lime and alumina is free quartz. Some
clays are almost destitute of felspar but comparatively
rich in mica, whilst others are the reverse, so that some
means of identifying the extraneous minerals is essential. When this is not used, the curious result is

obtained that German chemists calculate the alkalies, etc. to felspar whilst the French chemists, following Vogt, calculate them to mica; English ceramic chemists appear undecided as to which course to follow, and some of them occasionally report notable amounts of felspar in clays quite destitute of this mineral !

A statement of the composition of a 'clay' based on a mechanical separation of the coarser ingredients followed by an analysis of the finer ones and a calculation of the probable constituents of the latter, as already described, is known as a *proximate analysis* in order to distinguish it from an *ultimate analysis* which states the composition of the whole material in terms of its ultimate oxides.   A proximate analysis therefore shows the various materials entering into the composition of the clay in the following or similar terms :

Stones	per cent.
Gravel	,,
Coarse sand	,,
Medium sand	,,
Fine sand	,,
Silt	,,
Felspar or mica dust	,,
Silica dust	,,
'True clay[1]'	,,
Moisture	,,
Carbon	,,
Other volatile matter	,,

[1] In analytical reports a note should be appended stating that the figure under this term shows the proportion of the nearest approxima- tion to true clay at present attainable.

For some purposes it is necessary to show the proportion of calcium, iron and other compounds as in an ordinary ultimate analysis.

A comparison of the foregoing with an ultimate or 'ordinary' analysis of a clay (p. 16) will show at once the advantage of the former in increasing our knowledge of the essential constituent of all clays, if such a substance really exists. Its absolute existence is by no means proved, for, as will have been noticed, its composition is largely based on assumption even in the most thorough investigations, particularly of the admittedly less pure clays.

In the purer clays the problem is much simpler and in their case an answer of at least approximate accuracy can be given to the question 'What is clay?'

Even with these purer clays it is not sufficient to study an analysis showing the total amount of the silica, alumina and other oxides present; it is still necessary to effect some kind of separation into the various minerals of which they are composed. When, however, the accessory minerals do not exceed 5 per cent. of the total ingredients their influence is less important and the nature and characteristics of the 'clay substance' itself can be more accurately studied. By careful treatment of well selected china clays, for example, it is possible to obtain a material corresponding to the formula $Al_2O_32SiO_22H_2O$ within a total error of 1 per cent., the small amount of impurity being, as far as can be ascertained, composed of mica. So pure a specimen of clay is found on microscopical examination to consist of minute irregular grains of no definite form, together with a few crystals of the same composition and identifiable as the mineral ' kaolinite' (p. 107). This 'amorphous' material, which appears to be the chief constituent of all china clays and kaolins, has been termed *clayite* by Mellor (22).

Johnson and Blake, Aron and other observers have stated that the majority of the particles in china clays and kaolins are crystalline in form. Owing to their extreme smallness it is exceedingly difficult to prove that they are not so, though for all ordinary purposes they may be regarded as amorphous, the proportion of obviously crystalline matter present in British china clay of the highest qualities being so small as to be negligible.

Hickling (36), using an exceptionally powerful microscope, claims to have identified this 'amorphous' substance in china clay as ' worn and fragmental crystals of kaolinite,' and recently Mellor and Hold-croft and Rieke have shown that the apparently amorphous material shows the same endo- and exothermal reactions as crystalline kaolinite.

So far as china clays or kaolins are concerned, kaolinite or an amorphous substance of the same composition appears to be identical with the 'ideal clay' or 'true clay' whose characters have so long been sought.

This term—clayite—is very convenient when confined to china clays and kaolins, but it is scarcely legitimate to apply it, as has been suggested, to material in other clays until it has been isolated in a sufficiently pure form to enable its properties to be accurately studied.   This restriction is the more necessary as in one very important respect clayite obtained from china clay and some kaolins differs noticeably from the nearest approach to it obtainable from the more plastic clays : namely, in its very low plasticity.   This may be explained by the fact that it is only obtainable in a reasonably pure form in clays of a primary character, whilst the plastic clays have usually been transported over considerable areas and have been subjected to a variety of treatments which have had a marked effect on their physical character. Moreover, the fact that the purest 'clay' which can be isolated from plastic clays appears to be amorphous and to some extent colloidal greatly increases the difficulty of obtaining it in a pure state, especially as no liquid is known which will dissolve it without decomposing it.   The fact that it is not an elementary substance, but a complex compound of silica, alumina and the elements of water, also increases the intricacy of the problem, for these substances occur in other combinations in a variety of other minerals which are clearly distinct from clay.

Ever since the publication of Seger's memorable

papers (7), and to a small extent before that time, it has been generally understood that china clay or kaolin represented the true essential constituent of clays, but several investigators have been so imbued with the idea that all true clay substance must have a crystalline form that they have frequently used the term 'kaolinite' to include the 'amorphous' substance in plastic clays. This is unfortunate as it is by no means proved that the latter is identical with kaolinite, and a distinctive term would be of value in preventing confusion. Other investigators have used the word 'kaolin' with equal freeness, so that whilst it originally referred to material from a particular hill or ridge in China[1] it has now entered into general use for all clays whose composition approximates to that of china clay (p. 16) in which the plasticity is not well developed. Thus, in spite of the difference in origin between many German and French kaolins and the china clays of Cornwall, it is the custom in Europe generally to term all these materials 'kaolin.' Yet they are very different in many respects from the material originally imported from China.

As the essential clay substance has not yet been isolated in a pure form from the most widely spread plastic clays, but is largely hypothetical as far as they are concerned, the author prefers the term *pelinite*[2]

[1] *Kao-ling* is Chinese for a high ridge or hill.
[2] From the Greek πήλινος = made of clay.

when referring to that portion of any plastic clays
or mixtures of clays with other minerals which may
be regarded as being the constituent to which the
argillaceous portion of the material owes its chief
properties. In china clay and kaolin the 'true clay' is
identical with clayite—or even with kaolinite (p. 108)—
and there is great probability that this identity also
holds in the case of the more plastic clays of other
geological formations, but until it is established it
appears wisest to distinguish the hypothetical or
ideal clay common to all clays (if there is such a
substance) by different terms according to the extent to which its composition and characters of the
materials most closely resembling it are experiment-
ally known.

The substances most resembling this 'ideal clay'
which have, up to the present been isolated, are:

(*a*) *Kaolinite.* Found in a crystalline form in
china clays and kaolins (p. 107).

(*b*) *Clayite.* A material of the same chemical
composition as kaolinite, but whose crystalline nature
(if it be crystalline) has not been identified—chiefly
obtained from china clays and kaolins.

(*c*) *Pelinite.* A material similar to clayite, but
differing from it in being highly plastic and, to some
extent, of a colloidal nature—obtained from plastic
clays.

(*d*) *Laterite.* A material resembling clayite in

physical appearance, but containing free alumina and free silica (p. 80).

(*e*) *Clay Substance.* A general term indicating any of the foregoing or a mixture of them ; it is also applied (unwisely) to the material obtained when a natural clay is freed from its coarser impurities by elutriation (p. 7).

The Chief Characteristics of 'True Clay' from Different Sources.

In so far as it can be isolated *true clay* appears to be an amorphous, or practically amorphous, material which may under suitable conditions crystallize into rhombic plates of kaolinite. The particles of which it is composed are extremely small, being always less than 0·0004 in. in diameter. They adsorb dyes from solutions and show other properties characteristic of colloid substances though in a very variable degree, some clays appearing to contain a much larger proportion of colloidal matter than do others. To some extent the power of adsorption of salts and colouring matters appears to be connected with the plasticity (p. 41) of the material, but this latter property varies so greatly in clayite or pelinite from different sources as to make any generalization impossible.

True clay substance appears to be quite white, any colour present being almost invariably traceable

to ferric compounds or to carbonaceous matter. The
latter is of small importance to potters as it burns
away in the kiln. The specific gravity of clay substance is 2·65 according to Hecht, the lower figures
sometimes reported being too low. Its hardness is
usually less than that of talc—the softest substance on
Mohs' scale—but some shales are so indurated as to
scratch quartz. It is quite insoluble in water and in
dilute solutions of acids or alkalies, but is decomposed
by hydrofluoric acid and by concentrated sulphuric
acid when heated, alumina entering into solution and
silica being precipitated in a colloidal condition.

It absorbs water easily until a definite state of
saturation has been reached, after which it becomes
impervious unless the proportion of water is so large
and the time of exposure so great that the material
falls to an irregular mass which may be converted
into a slurry of uniform consistency by gently stirring
it.   With a moderate amount of water, pelinite
develops sufficient plasticity to enable it to be
modelled with facility, but clayite and some specimens
of pelinite are somewhat deficient in this respect. The
pelinitic particles usually possess the capacity to retain
their plasticity after being mixed with considerable
proportions of sand or other non-plastic material and
are then said to possess a high binding power (p. 28).

If a large proportion of water is added to a sample
of clayite or pelinite and the mixture is stirred into

a slurry it will be found to remain turbid for a considerable time and will not become perfectly clear even after the lapse of several days. Its power of remaining in suspension is much influenced by the presence of even small amounts of soluble salts in either the water or the clay substance, its precipitation being hastened by the addition of such salts as cause a partial coagulation of the colloidal matter present. Some specimens of clayite and pelinite retain their suspensibility even in the presence of salts, but this is only true of a very limited proportion of the substance. In most cases the presence of soluble salts causes the larger particles to sink somewhat rapidly and to carry the finer particles with them.

The rate at which a slip or 'cream' made of elutriated clay and water will flow through a small orifice is dependent on the viscosity of the liquid and this in turn depends on the amount of colloidal material present, *i.e.* on how much of the clay (pelinite) is in a colloidal form. Its viscosity is greatly affected by the addition or presence of small quantities of acid or alkali or of acidic or basic salts. Acids increase the viscosity; alkalies and basic salts, on the contrary, make the slip more fluid. Neutral salts behave in different ways according to the concentration of the solution and to the amount of clay (pelinite) present in the slip. If the slip contains so little water as to

be in the form of a thin paste, neutral salts usually
have but a small action, but when the slip contains
only a small proportion of clay (pelinite) the presence
of neutral salts will tend to cause the precipitation of
the clay.   In this way salts act in two quite different
directions according to the concentration of the slip.

On drying a paste made of clay and water the
volume gradually diminishes until the greater part
of the water has been removed ; after this the
remainder of the water may be driven off without any
further reduction in volume of the material.   This is
another characteristic common to colloidal substances
such as gelatin.   The material when drying attains
a leathery consistency which is at a maximum at the
moment when the shrinkage is about to cease ; on
further drying the material becomes harder and more
closely resembles stone.

Providing that wet clay is not heated to a
temperature higher than that of boiling water it
appears to undergo no chemical change and on cooling
it will again take up water[1] and be restored to its
original condition except in so far as its colloidal
nature may have been affected by the heating.   If,
however, the temperature is raised to about 500° C.
a decomposition of the material commences and water

[1] Some clays are highly hygroscopic and absorb moisture readily
from the atmosphere.   According to Seger (7) this hygroscopicity
distinguishes true clay from silt and dust.

is evolved. This water—which is commonly termed 'combined water'—is apparently an essential part of the clay-molecule and when once it has been removed the most important characteristics of the clay are destroyed and cannot be restored. The reactions which occur when clay is heated are complex and are rendered still more difficult to study by the apparent polymerization of the alumina formed. Mellor and Holdcroft (29) have recently investigated the heat reactions of the purest china clay obtainable and confirm Le Chatelier's view (10) that on heating to temperatures above 500° C. clay substance decomposes into free silica, free alumina and water, the two former undergoing a partial re-combination with formation of sillimanite ($Al_2O_3SiO_2$) if a temperature of 1200° C. is reached. Mellor and Holdcroft point out that there is no critical point of decomposition for clay substance obtained from china clay, as it appears to lose water at all temperatures, though its decomposition proceeds at so slow a rate below 400° C. as to be scarcely appreciable.

After the whole of the 'combined water' has been driven off, if the temperature continues to rise, it is found that at a temperature of 900° C. an evolution of heat occurs. This exothermal point, together with the endothermal one occurring at the temperature at which the decomposition of the clay seems to be most rapid, has been found by Le Chatelier, confirmed by

Mellor and Holdcroft, to be characteristic of clay substance derived from kaolin and china clay, and the two last-named investigators state that it serves as a means of distinguishing kaolinite or clayite from other alumino-silicates of similar composition. These thermal reactions have not, as yet, been fully studied in connection with plastic clays ; with china clay, as already noted, they probably indicate a polymerization of the alumina set free by the decomposition of the clay substance, as pure alumina from a variety of sources has been found by Mellor and Holdcroft to behave similarly.

On still further raising the temperature of pure clay (pelinite or clayite) no further reactions of importance occur, the material being practically infusible. If, however, any silica, lime, magnesia, alkalies, iron oxide or other material capable of combining with the alumina and silica is present as impurities in the clay substance, combination begins at temperatures above 900° C. This causes a reduction of the heat-resisting power of the material; the silicates and alumino-silicates produced fuse and begin to react on the remaining silica and alumina, first forming an impermeable mass in place of the porous one produced with pure clay substance, and gradually, as the material loses its shape, producing a molten slag if the 'clay' is sufficiently impure. As ordinary clays are never quite free from metallic compounds other than alumina, this formation of a fused portion—technically known as *vitrification* (p. 37)—occurs at temperatures depending on the nature of the materials present, so that a wide range of products is obtained, the series commencing with the entirely unfused pure clay (china clay), passing through the slightly vitrified fireclays, the more completely vitrified ball clays to the vitrifiable stoneware clays and ending with materials so rich in easily fusible matter as scarcely to be worthy of the name of clays.

The constitution of the clay molecule is a subject which has attracted the attention of many investigators and is being closely studied at the present time. It is a subject of peculiar difficulty owing to the inertness of clay substance at all but high temperatures, and to the complexity of reactions which take place as soon as any reagent is brought into active connection with it.

Without entering into details regarding the various graphic formulae which have been suggested, it is sufficient to state that the one which is most probably correct, as far as present knowledge goes, is Mellor's and Holdcroft's re-arrangement of Groth's formula(30)

$$\begin{array}{c} \text{HO}\diagdown \quad \diagup \text{OSiO}\diagdown \\ \quad \quad \text{Al}_2 \quad \quad \quad \text{O} \\ \text{HO}\diagup \quad \diagdown \text{OSiO}\diagup \\ \diagup \diagdown \\ \text{HO} \quad \text{OH} \end{array}$$

which on removal of the hydroxyl groups might be
expected to give the anhydride

$$\begin{array}{c} O \\ \diagdown \\ \diagup \\ O \end{array} Al_2 \begin{array}{c} OSiO \\ \diagup \\ \diagdown \\ OSiO \end{array} >$$

though in practice this substance—if formed at all—
appears to be instantly split up into Al_2O_3 and SiO_2.

By regarding the aluminium as a nucleus, as above,
and some aluminium silicates as hypothetical alumino-
silicic acids, as suggested by Ulffers, Scharizer,
Morozewicz (29) and others, clay substance may be
conveniently considered, along with analogous substances, as forming a special group quite distinct from
the ordinary silicates. In this way Mellor and Holdcroft (29) consider that clay substance is not a
hydrated aluminium silicate—as is usually stated in
the text-books—but an alumino-silicic acid, the salts
of which are the zeolites and related compounds.
From this hypothesis it naturally follows that clay
substance is analogous to colloidal silica which has
been formed by the decomposition of a silicate by
means of water and an acid.

If this view be correct, pure clay substance or true
clay is a tetra-basic alumino-silicic acid $H_4Al_2SiO_9$
or $Al_2Si_2O_5(OH_4)$. That its acid properties are not
readily recognizable at ordinary temperatures is due
to its inertness ; at higher temperatures its power of
combination with lime, soda potash and other bases

is well recognized, though the reactions which occur are often complicated by decompositions and molecular re-arrangements which occur in consequence of the elevated temperature.

There are a number of minerals which closely resemble clayite or pure clay substance in composition, the chief difference being in the proportion of water they evolve on being heated. Thus *Rectorite* $H_2Al_2Si_2O_8$, *Kaolinite* $H_4Al_2Si_2O_9$, *Halloysite* $H_6Al_2Si_2O_{10}$ and *Newtonite* $H_{10}Al_2Si_2O_{12}$. In the crystalline form these minerals may be distinguished from each other by means of the microscope, but as the chief materials of which clays are composed appears to be amorphous it is impossible to ascertain with certainty whether a given specimen of clay substance is composed of a mixture of these analogous minerals in an amorphous form or whether it consists entirely of clayite, *i.e.* the clay substance obtained from china clay. As already stated, the thermal reactions which occur on heating clayite appear to be characteristic of kaolinite whilst halloysite is completely decomposed at a temperature somewhat below 200° C.; but the not improbable presence of two or more of these alumino-silicic acids in clays of secondary or multary origin makes it almost impossible to determine whether clayite is an essential constituent of all clays or whether the purest clay substance (pelinite) obtained from some of the more plastic clays does

not possess a different chemical composition as well
as different physical properties.

The view that clays may. be regarded as impure
varieties of clayite is considered erroneous by
several investigators for various reasons.   For
instance, felspar is rarely found in china clays, but
is a common constituent of secondary (plastic) clays.
J. M. van Bemmelen (26), who has found that the
alumina-silica ratio of clays produced by weathering
is always higher than that in clays of the china clay
type produced by hypogenic action.   In a number of
clays examined he found that a portion was soluble
in boiling hydrochloric acid whereas clayite is scarcely
affected by this treatment.   He also found a varying
proportion of alumino-silicate insoluble in hydrochloric
acid but dissolved on treatment with boiling sulphuric
acid and subsequently with caustic soda solution ;
this latter he considers to be true clayite.   Unfortunately, his results were obtained by treating
the crude clay with acid, instead of first removing
such non-plastic materials as can be separated by
washing, so that all that they show is that some clays
contain alumino-silicates of a nature distinct from
clayite in addition to any clayite which may be found
in them.

The fact that all clays when heated to 700 or 800° C.
readily react with lime-water to form the same
calcium silicates and aluminates indicates so close a

resemblance between the clay substance obtainable from different sources as to constitute strong evidence of the identity of this substance with clayite or with materials so analogous to it as to be indistinguishable from it under present conditions.

In all probability, the plastic clays have been derived from a somewhat greater variety of minerals than the primary clays (p. 71) and under conditions of decomposition which differ in details, though broadly of the same nature as those producing china clays. The presence of colloidal matter suggests a more vigorous action—or even a precipitation from solution—instead of the slower reactions which result in the formation of the kaolinite crystals.

The much smaller particles present in plastic clays also indicate a more complete grinding during the transportation of the material or some form of precipitation. If, as Hickling suggests, all clays are direct products of the decomposition of *mica*, the fact that several varieties of mica are known and that the conditions under which these decompose must vary considerably, afford a good, if incomplete, explanation of some of the widely diverse characteristics observed in different clays.

Notwithstanding the great complexities of the whole subject and the apparently contradictory evidence concerning some clays, there is a wide-spread feeling that whatever may be the mineral from which a given clay has been derived, the *true clay substance*, which is its essential constituent, would (if it could be isolated in a pure state) prove to be of the same composition as kaolinite obtainable from china clay of exceptional purity. The purest clay substances (pelinite) yet obtained from some of the most plastic clays are, however, so impure as to make any detailed investigation of their composition by present methods abortive. The methods of synthesis which have proved so successful in organic chemistry have hitherto yielded few intelligible results with clays, on account of the complexity of the accessory reactions which occur.

THE DIFFERENCE BETWEEN PURE CLAY SUBSTANCE AND ORDINARY CLAYS.

The properties and characteristics of *true clay* are very seriously modified by other materials which may be associated with it. This may be perceived by comparing the properties of clays mentioned in Chapter I with those of various forms of true clay just given. Moreover, as true clay never occurs in a perfectly pure state in nature, the properties of clays must be largely dependent on the accessory ingredients.

Silica, for example, when alone is a highly refractory material, but in the presence of true clay it reduces the refractoriness of the latter. Lime has a similar effect though its chemical action on the clay is entirely different. A very small proportion of some substances—notably the oxides of sodium and potassium—will greatly alter the behaviour of true clay when heated and will produce an impervious mass in place of a porous one.

For these reasons, it is necessary in studying clays to pay attention to both their physical and chemical properties and to separate the material into fractions so that each of these may be studied separately and their individual as well as their collective characteristics ascertained. Failure to do this has been the cause of much obscurity and confusion in investigations on certain clays composed of a considerable proportion of non-argillaceous material which ought to have been separated before any attempt was made to study the true clay present.

There is, therefore, a considerable difference between a natural clay and the pure clay substance theoretically obtainable from it ; this difference being most marked in the case of low-grade brick clays of glacial origin, which may contain 50 per cent. or more of adventitious materials. If used in a natural state they would be found to be valueless on account of their impurities giving them characteristics of a highly undesirable character, whereas the true clay in them is found—in so far as it can be separated—to bear a close resemblance to that obtained from a high grade, plastic, pottery clay. Unfortunately, it is, at present, impossible to isolate this clay substance in anything approaching a pure form, and many clays are without commercial value because of comparatively small proportions of impurities which cannot be separated from the clay substance without destroying the latter.

CLASSIFICATION OF CLAYS.

Owing to the widely differing substances from which clays can, apparently, be formed and the peculiar difficulties which are experienced in investigating the nature of clay substance from different sources, it is by no means easy to devise a scheme of classification of clays, though many of these have been attempted by different scientists.

The classification adopted by geologists is based on the fossil remains and on the stratigraphical position of clays relative to other rocks, as described in Chapter II. This is of great value for some purposes, but the composition of the substances termed 'clay' by geologists differs so greatly, even when only one formation is considered, as to make their classification of little or no use where the value or worthlessness of the material depends upon its composition. Thus the so-called Oxford clay ranges from a hard silicious shale to a comparatively pure clay; some portions of it are so contaminated with calcareous and ferruginous matter as to make the material quite useless for the potter or clayworker. A geological classification of clays is chiefly of value as indicating probable origins, impurities and certain physical properties; but the limits of composition and general characteristics are so wide as to make it of very limited usefulness.

The classification of clays on a basis of chemical composition is rendered of comparatively little value by the large number of clays which occupy ill-defined borders between the more clearly marked classes. Moreover, attempts to predict the value and uses of clays from their chemical composition are generally so misleading as to be worse than useless, unless a knowledge of some of the physical characters of the clays is available. It is, of course, possible to differentiate some clays from others by their composition, but not with sufficient accuracy to permit of definite and accurate classification.

A classification based exclusively on the composition of clays is equally unsatisfactory for other reasons, the chief of which is the placing together of clays of widely differing physical character, and the separation of clays capable of being used for a particular purpose. To some extent the latter objection may be disregarded, though it is of great importance in considering the commercial value of a clay.

Classification based on the uses of clays of different kinds has been suggested by several eminent ceramists, but is obviously unsatisfactory, particularly as it is by no means uncommon to use mixtures of clays and other minerals for some purposes. Thus stoneware clays must be vitrifiable under conditions which may be defined with sufficient accuracy, but many manufacturers of stoneware do not use clays which are naturally vitrifiable ; they employ a mixture of refractory clay and other minerals to obtain the material they require.

A classification based on the origin of clays regarded from the petrological point of view offers some advantages, but is too cumbersome for ordinary purposes and suffers from the disadvantage that the origin of some important clays is by no means clearly known.

The author prefers a modification of Grimsley's and Grout's classification (31) as follows :

I.   Primary clays.

(a)   Clays produced by 'weathering' silicates—as some kaolins.

(b)   Clays produced by lateritic action—very rich in alumina, some of which is apparently in a free state.

(c)   Clays produced by telluric water containing active gases (hypogenically formed clays)—as Cornish china clay.

II.   Secondary clays.

(*d*)   Refractory[1] secondary clays—as fireclays and some pipe clays.

(*e*)   Pale-burning non-refractory clays—as pottery clays, ball clays and some shales.

(*f*)   Vitrifiable clays—as stoneware clays, paving brick clays.

(*g*)   Red-burning and non-refractory clays—as brick and terra-cotta clays and shales.

(*h*)   Calcareous clays or marls, including all clays containing more than 5 per cent. of calcium carbonate.

III.   Residual clays.

(*i*)   Clays which have been formed by one of the foregoing actions and have been deposited along with calcareous or other matter but, on the latter being removed by subsequent solution, the clay has remained behind—as the white clays of the Derbyshire hills.

Some further subdivision is necessary for special purposes, particularly in sections *e*, *f* and *h*, but to include further details would only obscure the general scheme. Some clays will, apparently, be capable of classification in more than one section, thus a vitrifiable clay may owe its characteristic to a high proportion of calcium carbonate and so be capable of inclusion as a calcareous clay. Broadly speaking, however, if the clay is tested as to its inclusion in each section of the scheme in turn it will be found that its

---

[1] A refractory clay is one which does not soften sufficiently to commence losing its shape at any temperature below that needed to bend Seger Cone 26 (approximately 1600° C.) (see p. 116).

highest value will be in the section which is nearest to the first in which the clay can legitimately be placed.

From a consideration of a classification such as the foregoing, together with a detailed study of the physical and chemical properties of the material as a whole, and also of the various portions into which it may be divided—particularly that which has been isolated by mechanical methods of purification and separation—it is not difficult to gain a fairly accurate idea of the nature of any clay. Although the present state of knowledge does not permit them to be classified with such detail as is the case with plants, animals, or simple chemical compounds, the study of clays and the allied materials has a fascination peculiarly its own, not the least interesting features of which are those properties of the clay after it has been made into articles of use or ornament. These are, however, beyond the scope of what is commonly understood by the term 'the natural history of clay.'

BIBLIOGRAPHY

A complete bibliography of clay would occupy several volumes. The following list only includes the more accessible of the works quoted in the text.

1. "Second Report of the Committee on Technical Investigation—Rôle of Iron in Burning Clays." Orton and Griffith. Indianopolis. 1905.
2. "British Clays, Shales and Sands." Alfred B. Searle. Charles Griffin and Co. Ltd. London. 1911.
3. "Transactions of the English Ceramic Society." v. p. 72. Hughes and Harber. Longton, Staffs. 1905.
4. "Royal Agricultural Society's Journal." xi.
5. "Die Tone." P. Rohland. Hartleben's Verlag. Vienna. 1909.
6. "Clays: their Occurrence, Properties and Uses." H. Ries. Chapman and Hall. London. 1908.
7. "Gesammelte Schriften." H. Seger. Tonindustrie Zeitung Verlag. Berlin. 1908.
8. "Tonindustrie Zeitung." 1902. p. 1064.
9. "Tonindustrie Zeitung." 1904. p. 773.
10. "Treatise on Ceramic Industries." E. Bourry (Revised translation by A. B. Searle). Scott, Greenwood and Son. London. 1911.
11. "The Colloid Matter of Clay." H. E. Ashley. U.S.A. Geological Survey Bulletin 388. Washington. 1909.
12. "Sprechsaal." 1905. p. 123.
13. "Action of Heat on Refractory Materials." J. W. Mellor and F. J. Austen. Trans. Eng. Cer. Soc. vi. Hughes and Harber. Longton, Staffs. 1906.
14. "Wiedermann's Annalen." vii. p. 337.
15. "Geological Contemporaneity." 1862.
16. "Geological Magazine." iv. pp. 241, 299.
17. "La Céramique industrielle." A. Granger. Gauthier Frères. Paris. 1905.

18. "American Journal of Science." 1871. p. 180.
19. "The Hensbarrow District." J. H. Collins. Geological Survey. 1878.
20. "Monographs of the U.S.A. Geological Survey." XXVIII. C. R. van Hise. 1897.
21. "On Kaolinite and Pholerite." American Journal of Science. XLIII. 1867.
22. "The Nomenclature of Clays." J. W. Mellor. Eng. Cer. Soc. VIII. Hughes and Harber. Longton, Staffs. 1908.
23. "On the present distribution of Coal Balls." M. C. Stopes and D. M. S. Watson. Phil. Trans. Royal Society. B. Vol. CC. 1908.
24. "Natural History of Coal." E. A. N. Arber. Cambridge University Press. 1911.
25. "Modern Brickmaking." A. B. Searle. Scott, Greenwood and Son. London. 1911.
26. "Die verschiedene Arten der Verwitterung." J. M. van Bemmelen. Zeits. angewandte Chemie. LXVI. Leopold Voss Verlag. Hamburg. 1910.
27. "Pyrometrische Beleuchtung." Carl Bischof. Tonindustrie Zeitung. 1877.
28. "Die feuerfeste Tone." Carl Bischof. Quandt and Haendler. Leipzig. 1904.
29. "The Chemical Constitution of the Kaolinite Molecule." Trans. Eng. Cer. Soc. X. Hughes and Harber. Longton, Staffs. 1911.
30. "Tabellarische Uebersicht der Mineralien." P. Groth. Brunswick. 1898.
31. "West Virginia Geological Survey." III. 1906.
32. "Memoirs of the Geological Survey." London.
33. "The Publications of Stanford's Geographical Institute." London.
34. "Handbuch der gesam. Tonwarenindustrie." B. Kerl. Verlag der Tonindustrie Zeitung. 1910.
35. "Causal Geology." E. H. L. Schwarz. Blackie and Sons, Ltd. 1910.
36. "China Clay: its nature and origin." G. Hickling. Trans. Inst. Mining Engineers. 1908.

INDEX

Absorption, 40, 151
Absorptive power of clays, 40
Accumulation of clays, 84
Acid-proof ware, 113
Acids, effect of, 106, 151, 152
Adsorption, 40, 150
Agriculture, clays in, 5, 56, 57, 59, 61, 62, 63, 67
Air, 43, 85
Alkalies in clay, 38, 115, 133, 142, 143, 155
Alluvial deposits, 68, 87, 112, 132
Alum clays and shales, 57, 123, 124
Alum manufacture, clays for, 124
Alumina, 6
Alumina, free, 80, 82, 154
Alumina-silica ratio, 133, 159
Alumino-silicic acid, 6, 76, 81, 118, 155, 157
Aluminous clays, 82, 117
'Amorphous' clay, 107, 146
Analyses of clays, 16, 141, 144
Anauxite, 134
Architectural ware, 129, 130
Argillaceous earths, 1
Argillaceous limestone, 88, 132
Associated rocks, 48

Bagshot clays and sands, 64, 125
Ball clays, 6, 19, 28, 62, 64, 82, 110, 115, 119, 125, 138, 141, 156, 166

Bending of clay, 33
Bibliography, 168
Binding power, 28, 151
Binds, 53
Bituminous shales, 57, 59
Black spots, 14, 128
Black ware, 113
Bleaching oil, 134
Blue bricks, 13, 56
Bone-ash, 110
Boulder clays, 3, 7, 10, 65, 101
Bovey Tracey clay, 62
Brick clays, earths and shales, 1, 2, 5, 10, 11, 12, 13, 31, 37, 40, 46, 56, 57, 59, 61, 63, 65, 67, 68, 91, 100, 101, 104, 112, 117, 123, 125, 129, 138, 162, 166
Brittleness, 46
Brown ware, 113
Buff bricks, 128
Burned clay, 28, 31, 41, 119, 121

Calcareous clays, 38, 61, 68, 88, 133, 139, 166
Calcareous sands, 88
Calcium, see *Lime compounds*
Cambrian clays, 51
Carbon in clay, 15, 119, 144
Carbonates in clay, 10, 82
Carboniferous clays and shales, 52, 124
Carboniferous limestone, 52, 108

Carclazite, 78, 106
Cellulose in clays, 27
Cement clays, 57, 104, 131
Chalcopyrite, 14
Chalk, 10, 11, 61, 67, 68, 88, 116, 127, 128, 132, 134, 139
Chamotte, 121
Chemical properties of clay, 6
China clay rock, 78, 106, 116
China clays, 2, 5, 6, 7, 9, 22, 27, 40, 49, 64, 71, 75, 78, 82, 84, 104, 110, 116, 141, 146, 147, 148, 156, 165
China-ware, 109, 110
Chinese clay, 73
Classification of clays, 163
Clay molecule, 156
Clay-shales, 122
Clay substance, 135 *et seq*.
Clay substance, defined, 150
Clayite, 83, 107, 147, 149
Clinker, 132
Clunches, 118
Coagulated clays, 97
Coagulation, 43, 152
Coal Measure clays and shales, 53, 96, 103, 117, 124, 130
Coarse pottery, 112
Cobalt, 110
Colloid theory, 97
Colloidal properties of clay, 25, 81, 82, 97, 106, 147, 150, 152
Colloidal silica, 81, 134, 157
Colloids, 24, 41, 43, 76, 160
Colluvial clays, 99
Colours of burned ware, 19, 123, 131
Colours of clays and shales, 19, 59, 115, 119, 124, 126
Combined water, 45, 154
Common clays, 3
Composition of clays, 4, 6, 16, 23,

35, 44, 107, 117, 118, 133, 134, 144, 156, 164
Composition of clays(burned), 46
Cornish stone, 110, 116
Cracked ware, 46, 127, 130, 131
'Cream,' 39, 43, 152
Cretaceous clays, 61
'Crumb' of clay, 24
Crushing clay, 45
'Crystalline' clay, 107, 146, 148
Crystals in clay-ware, 46

Decantation, 139
Decomposition of clay, 154
Definitions of clay, 2–5, 120, 135, 149, 150
De-greasing wool, 134
Deposition of clays, 49, 51, 90, 99
Devonian clays, 51
Diluvial clays, 99
Dinas rock, 54
Disintegration, 102
Distribution of clays, 1
Drain-pipe clays, 112, 113
Drift, 65, 101
Drift clays, 101
Drying clays, 27, 127, 153
Durability, 131
Dyes, 41, 150

Earth movements, 85, 96
Earthenware, 37, 112
Earths for bricks, see *Brick clays*
Electrolytes, 43
Elutriation, 8, 137, 140
Eocene clays, 63
Epigenic clays, 82
Erosion, 89, 99, 100
Estuarine clays, 90, 93, 118
Etruria marls, 55, 130
Expansion, 32
Exposure, 43

Faience, 129
Farewell Rock, 54
Fat clays, 29
Felspar, 7, 8, 41, 74, 104, 110, 116, 141, 144, 159
Ferric and Ferrous compounds, 12, 121, see *Iron*
Fine clays, 112
Fineness, see *Texture*
Firebricks, 14, 54, 61, 116
Fireclay, 33, 35, 52, 54, 82, 104, 108, 116, 123, 156, 166
Fissile clays, 117
Flint, 110, 116
Flint clays, 117
Floods, 85, 87, 99
Flower-pot clays, 57, 110
Fluoric vapours, 75, 77, 165
Fluviatile clays, 88, 92
Fluxes, 8, 11, 38, 39, 115, 116, 131
Food-clays, 1
Formation of clays, 48, 70
Formula of clay, 156
Free alumina, 80, 154
Free silica, 7, 80, 154, 161
Frost, 43, 86
Fuller's earth, 59, 133
Fulling cloth, 1, 133
Fusibility, 32, 58, 113, 116
Fusible clays, 116
Fusing point, 32
Fusion, 47, 113, 120, 132, 155

Ganister, 52, 54, 118
Gault, 61, 132
Geological classification, 163
Geological nature of clay, 4, 50
Glacial clays, 65, 100, 162
Glaciers, 85, 89, 100
Glass, 116
Glassy structure, 47

Glazed bricks, 119
Glazed pottery, 129
Glazed terra-cotta, 56, 129
Grades of fireclay, 120
Gravel, 7, 62, 65, 89, 100, 101, 102, 138, 144
Green colour, 14
Greensand, 133
Grinding, 80, 121
Grit, 112, see also *Millstone Grit*
Grog, 28, 31, 41, 119, 121
Growan, 78
Gypsum, 10, 12, 62

Halloysite, 118, 158
Hardness, 45
Heat, effects of, 28, 37, 39, 45, 80, 122, 146, 153, 154, 158, 159
Hydrargillite, 80
Hydro-alumino-silicates, 6
Hydrocarbons in clay, 15
Hydrolysis, 78, 97
Hygroscopic clays, 153
Hypogenic clays, 165

Ice-action, 85, 100
Ideal clay, 146
Impermeability, 40, 113
Impervious articles, 113, 155
Impurities in clays, 7, 49, 82, 102, 104, 109, 121, 126, 142, 143, 155, 162, 163
Ions, 43
Indurated clays, 18
Infusibility, 106, 119, see *Refractoriness*
Iron compounds, 7, 10, 12, 13, 20, 62, 112, 119, 121, 128, 133, 141, 145, 164
Ironstone, 62
Irregularity in shape, 131

Jurassic clays and shales, 57, 124

Kao-ling, 148
Kaolinite, 19, 80, 105, 107, 146, 149, 158
Kaolinization, 76, 77, 79
Kaolins, 9, 21, 49, 64, 71, 73, 76, 79, 82, 84, 104, 116, 141, 146, 147, 148, 165
Keele series, 55
Kellaways clay, 59, 61
Keuper marls, 57
Kiln shrinkage, 30
Kimeridge clays, 59
Knotts, 124

Lacustrine clays, 90, 91
Lake-deposited clays, 85, 88, 91
Lakes, 85, 88
Laminated clays, 53, 117, 122
Laterite, 80, 149
Lateritic action, 80, 165
Lateritic clays, 82
Lean clays, 29
Liassic clays and shales, 57, 125, 132
Lime, 7, 10, 102, 159
Lime compounds, 10, 11, 38, 41, 47, 113, 115, 116, 121, 127, 139, 142, 143, 145, 155, 157, 162, 164
Limestone, 10, 11, 52, 59, 61, 62, 88, 102, 117, 127, 132, 139
Lime troubles, 11
Loam, 57, 67, 88
London clay, 62, 63, 125
Ludwig's chart, 35

Magnesium compounds, 7, 10, 11, 41, 47, 113, 115, 116, 121, 133, 155
Malm-bricks, 11

Malms, 10, 68
Marcazite, 10, 13
Marine clays, 61, 93
Marls, 10, 51, 54, 55, 57, 67, 68, 88, 130, 132, 166
Mechanical analysis, 137
Medway mud, 132
Melting point, 31, 32
Mica, 7, 8, 76, 104, 105, 116, 140, 141, 144, 160
Microscopical examination, 18, 105, 143, 158
Millstone grit, 54, 55, 117
Mineral nature of clay, 3
Minerals resembling clay, 158
Mining ball clay, 111
Modelling clays, 130
Moisture, 15, 144
Molecular attraction, 22
Molecular constitution of clay, 21, 156
Montmorillonite, 134
Mundic, 13
Muscovite, 105

Newtonite, 158
Nodules, 121, 127
Non-plastic material, 43, 121, 151
Non-refractory clays, 166

Occurrence of clays, 48, 116
Ocean currents, action of, 89
Odour of clay, 19
Oil, bleaching, 134
Oil shales, 61, 122, 123
Oolite clays, 59, 134
Ooze, 95, 99
Organic matter, 19, 119
Origins of clays, 48, 71, 160, 165
Oxford clay, 59, 95, 164
Oxides in clay, 10, 82

Paint, clays for, 109
Paper, clays for, 109
Particles, nature of, 18, 31, 106, 107, 150
Paving brick clays, 166
Pelagic ooze, 95, 99
Pelinite, 83, 148, 149
Permian clays and shales, 57, 112, 124
Pholerite, 117
Physical characters of clays, 17
Picking clay, 121
Pipe clays, 64, 65, 82, 109, 166
Plant-extracts in clays, 26
Plastic clays, 2, 43, 65, 67, 82, 88, 102, 112, 123, 147, 148, 160
Plasticity, 20–27, 41, 46, 97, 98, 99, 108, 109, 112, 117, 123, 125, 127, 151, 160
Pleistocene clays, 67
Pockets, 65, 85, 101, 116, 166
Porcelain, 37, 46, 73, 109, 110, 125, 129
Pores in clay, 30, 114
Porosity, 30, 39, 121, 131, 155
Portland cement, 131
Potash compounds, 7, 10, 113, 115, 116, 121, 157, see *Alkalies*
Pottery clays, 1, 5, 31, 46, 66, 100, 101, 104, 110, 112, 114, 125, 129, 162, 166
Precambrian clays, 51
Precipitated clays, 97, 152
Primary clays, 70, 71, 84, 165
Proximate analysis, 16, 144
Purbeck clays, 59
Pure clays, 5, 6, 7, 142, 155, 156
Purification of clay, 7, 66, 78, 104, 113, 128, 140
Pyrites, 10, 13, 44, 56, 57, 119, 124, 128

Quartz, 8, 104, 110, 118, 140, 141, 143

Rain, 44, 85, 86
Rational analysis, 141
Reading clays, 63
Recent clays, 67
Rectorite, 158
Re-deposited clays, 98
Red bricks, 12
Red burning clays, 141, 142, 166
Red iron oxide, 12
Red ware, 113
Reduction in volume, 30
Refractoriness, 34, 119, 120, 123, 155
Refractory articles, 5, 119
Refractory clays, 9, 32, 33, 35, 38, 52, 65, 82, 104, 116, 165, 166
Residual clays, 70, 84, 166
Resistance to abrasion, 119
Resistance to corrosion, 119
Resistance to crushing, 46
Resistance to cutting, 119
Resistance to temperature, see *Refractoriness*
Resistance to weathering, 76
Ringing sound, 110
River-deposited clays, 88
Rivers, 85, 87
Rock binds, 53
Rockingham, 113
Rock-like clays, 2
Rocks associated with clay, 48
Roman cements, 133
Roofing tiles, 57, 63, 126, 128

Sagger marls, 54
Sand, 7, 31, 41, 62, 82, 89, 100, 101, 117, 133, 138, 144
Sandstones, 53
Sandy clays, 68

Sandy loams, 88
Sandy marls, 88
Sanitary articles, 5, 113
Sawdust, 40
Scum, 10
Sea, action of, 85, 89, 99
Sea-deposited clays, 93
Secondary clays, 70, 82, 83, 166
Sedimentary rocks, 48
Sedimentation of clay, 43, 88, 90, 104
Seger cones, 33, 34
Selection of clay, 122
Selenite, 10
Separation of clays, 90, 145
Settling, 43
Sewerage pipes, 113
Shale oil, 15, 61
Shale tar, 123
Shales, 2, 5, 51, 52, 53, 57, 61, 96, 104, 122, 130, 132, 138, 151, 162, 166
Shrinkage, 11, 29, 58, 68, 102, 110, 117, 119, 121, 127, 131, 153
Sifting, 7, 138
Silica, 6, 7, 80, 154, 155, 161
Silica rock, 118
Silicates, 8, 82
Siliceous clays, 117
Sillimanite, 46, 154
Silt, 90, 91, 99, 139, 144
Silurian clays and shales, 51, 124
Sintering, 38
Size of particles, 18, 21, 31, 106, 107, 150
' Skeleton,' 115
' Skin' on ware, 131
Slag in bricks, etc., 11, 13, 119, 155
Slates, 51
Slurry, 39, 43, 152

Snow, 85
Soda compounds, 7, 10, 113, 115, 116, 121, 157, see *Alkalies*
Softening point, 33
Soil, see *Agriculture*
Solubility of clay, 151, 159
Soluble salts, 10
Sorting, 90
Sources of clays, 85
Specification of fire clays, 120
Specific gravity, 18, 106, 151
Staffordshire bricks, 13, 56
Standard clay, 4
Stone, Cornish, 110, 116
Stoneware clays, 104, 112, 113, 156, 165, 166
Stones, 7, 65, 100, 102, 128, 138, 144
Streams, 85, 86
Strength, 23, 45, 113
Sub-surface clays, 5
Sulphates in clay, 10, 12, 82
Sulphides in clay, 10, 82
Sulphuric acid, 124
Sunlight, 45
Surface clays, 2, 5, 52, 112
Suspension of clay, 43, 90, 140, 152
Swelling, 15, 102

Tannin in clay, 25, 26, 41
Telluric water, 165
Temperature, resistance to, 119, 120
Tensile strength, 23, 45
Terra-cotta clays and shales, 5, 10, 12, 31, 46, 56, 63, 91, 104, 123, 124, 129, 166
Tertiary clays, 62
Texture, 112, 130
Thermal reactions, 146, 154, 158
Tiles, 1, 5, 57, 91, 101, 129

Titanium compounds, 121
Tourmaline, 76, 104
Transportation of clays, 49, 86, 98, 99, 100
Transported clays, 70
Triassic clays, 57, 112, 130
True clay, 144, 146, 149, 150, 160
Twisted ware, 114, 129, 131
Types of clay, 82

Ultimate analysis, 16, 144
Ultra-marine, clays for, 109
Underclays, 53, 117, 118
Uses of clay, 1, 165

Valuation of clay, 103, 109, 123, 126, 162, 165
Vegetable matter, 15, 119
Veins, 85
Viscosity, 152
Vitrifiable clays, 113, 156, 166

Vitrification, 15, 20, 37, 112, 113, 114, 156
Vitrification range, 38, 114, 115, 116, 129
Volcanoes, 85

Warp, 99
Warped ware, 114, 129, 131
Washing, 7, 79
Water, effect of, 74, 76, 81, 85, 86, 151
Water in clays, 15, 17, 29, 39, 45, 154
Wealden clay, 62
Weathering, 44, 74, 76, 79, 80, 97, 107, 165
White bricks, 68, 128
White clays, 70, 166
Wind, 86

Zeolites, 157

For EU product safety concerns, contact us at Calle de José Abascal, 56–1°, 28003 Madrid, Spain or eugpsr@cambridge.org.

www.ingramcontent.com/pod-product-compliance
Ingram Content Group UK Ltd.
Pitfield, Milton Keynes, MK11 3LW, UK
UKHW010850090126
466816UK00011B/142